THE SPONTANEOUS GENERATION CONTROVERSY

FROM DESCARTES TO OPARIN

THE SPONTANEOUS GENERATION CONTROVERSY FROM DESCARTES TO OPARIN

JOHN FARLEY

THE JOHNS HOPKINS UNIVERSITY PRESS

BALTIMORE AND LONDON

This book has been brought to publication with the generous assistance of the Andrew W. Mellon Foundation.

The Johns Hopkins University Press, Baltimore, Maryland 21218
The Johns Hopkins Press Ltd., London

Library of Congress Catalog Card Number 76-47379
ISBN 0-8018-1902-4

Library of Congress Cataloging in Publication data will be found on the last printed page of this book.

To my wife
Grace
and our children
Gael, Gyneth, James, and Gilmour
this book is happily dedicated

CONTENTS

ACKNOWLEDGMENTS

Must living beings necessarily arise from parents? The answer to that question embroiled natural philosophers in a long and often bitter argument extending over three hundred years. My interest in the controversy began during a sabbatical year in the Department of the History of Science at Harvard University, where I discovered that my beloved parasitic worms lay at the center of the storm. Yet Louis Pasteur breathed no word of this! I cannot begin to express my deep appreciation to Everett Mendelsohn, not only for inviting me to spend that year in Cambridge, but also for encouraging me to use the facilities of the department and of the magnificent Harvard libraries on numerous occasions since that time. To him and other members of the department my sincere thanks for their interest and hospitality.

My thanks go also to many individuals whose advice was always so freely given. To John Engroff, Carol Adams, Philip Lawrence, Richard Burkhardt, and Frederic Gregory, I am especially grateful. I owe a special debt of gratitude to Gerald Geison for allowing me to utilize our article on the Pasteur-Pouchet debate, which appeared in the *Bulletin of the History of Medicine* in 1974 (Volume 48), and for his very helpful insights into the work of Pasteur. In addition I must thank Dorian Kottler for making it possible for me to read her dissertation on Pasteur and molecular dissymmetry. I also thank the Reidel Publishing Company for permission to quote from two articles on the spontaneous generation controversy, which I wrote for the *Journal of the History of Biology*.

I have benefited greatly from the criticisms of those who have read earlier drafts of the manuscript. Particularly warm thanks go to Frederic L. Holmes who read over the very first draft during a delightful visit to London, Ontario. I appreciate also the very perceptive criticisms offered by William Provine and Donna Haraway.

I wish also to thank Carol Cooper and Elaine Robart for typing the manuscript; Mary Primrose, Derek Sarty, and Goldie Gibson for their assistance in art work and photography; and Barry Fox for his help with things pertaining to the English language.

I acknowledge with gratitude the financial assistance of The

x

ACKNOWLEDGMENTS

Canada Council which enabled me to work in Cambridge on many occasions over the past few years.

Finally, I would like to offer especially warm thanks to Vice-President Andrew MacKay of Dalhousie University and to Kraft von Maltzahn and my other colleagues in the Biology Department. They not only tolerated a renegade in their midst, but actually encouraged his pursuits in a field on which many scientists cast a jaundiced eye. Without their support I doubt whether this book would have been completed.

Halifax, Nova Scotia   JOHN FARLEY

And God said, Let the earth bring forth grass, the herb yielding seed, and the fruit tree yielding fruit after his kind, whose seed is in itself, upon the earth: and it was so. . . . And the evening and the morning were the third day.

And God said, Let the waters bring forth abundantly the moving creature that hath life, and fowl that may fly above the earth in the open firmament of heaven. . . . And the evening and the morning were the fifth day.

And God said, Let the earth bring forth the living creature after his kind, cattle, and creeping thing, and beast of the earth after his kind: and it was so. . . .

And God saw everything that he had made, and, behold, it was very good. And the evening and the morning were the sixth day.

Thus the heavens and the earth were finished, and all the host of them.

And on the seventh day God ended his work which he had made. . . .

GENESIS 1 AND 2

THE SPONTANEOUS GENERATION CONTROVERSY

FROM DESCARTES TO OPARIN

ONE

INTRODUCTION

The problem of spontaneous generation, or equivocal generation as it was sometimes called,[1] occupied the attention of many individuals living in different countries at different times and adhering to a wide spectrum of philosophical and theological views. Thus, it is only to be expected that such individuals would not define what they believed in exactly the same way. Basically, however, all proponents of spontaneous generation believe that *some living entities may arise suddenly by chance from matter independently of any parent.* There was no uniformity, however, in their concept of what constituted the nature of these living entities and of the nature of the matter from which they arose. Neither would they agree on the time at which spontaneous generation could occur nor on the precise implications of the term *by chance.*

Historically, many types of living entities have been viewed as potential products of spontaneous generation. They ranged from fully formed organisms such as algae, fungi, insects, parasitic worms, infusorians, and bacteria, to "living matter" and "living molecules." Such entities were thought to arise either from inorganic matter (*abiogenesis*) or from organic matter which was itself alive or derived from a living organism (*heterogenesis*). This distinction between abiogenesis and heterogenesis became crucial at times when organic matter was thought to be produced only from living organisms. At such times "vitalists," who would necessarily deny any possibility of abiogenesis, could readily accept heterogenesis. Proponents of spontaneous generation also differed over whether it was a recurring phenomenon or one which had taken place only in the past.

Another significant variable in the controversy concerned the meaning of the term, *by chance.* In the eighteenth and much of the nineteenth century, *chance* implied an unlawful, accidental, unknowable event, in antithesis to the universality of strictly deterministic natural laws. Gradually the realization grew that, although single events may not be determinable, in aggregate they do exhibit certain regularities. Such *chance* events may thus be predicted on the basis of the probability of their occurrence during a certain specified time interval. Rather than viewing natural phenomena in terms of being either *by chance* or *by law, all* events can be placed along a continuum of probability—those events seen to be lawful having a probability approaching one, those seen to be by chance having a probability closer to zero. In all cases the time interval must be specified.

Such historical variations in the meaning of spontaneous generation have been ignored in traditional accounts of the controversy. This is not surprising, for until fairly recently writings in the history of science have had as their objective a desire to deepen understanding of contemporary science. They narrated how, when and by whom modern concepts were established through a slow, but progressive, evolution to the present. In the words of T. S. Kuhn, "observations, laws, or theories which contemporary science had set aside as error or irrelevancy were seldom considered unless they pointed a methodological moral or explained a prolonged period of apparent sterility."[2] Basic to this "Whig historiography" is the assumption that there exists an absolute scientific truth, and that historians of science must narrate the discovery of this truth. As a corollary to this, scientists who erred were often examined in an attempt to elucidate why they were so led astray.

Modern historians of science, of course, have abandoned such practices. Arguments over the merits and faults of Whig historiography no longer occupy their attention. Nevertheless, many episodes in the history of science are still seen in Whiggish terms, among them being the standard accounts of the spontaneous generation controversy. Indeed, this controversy provides a paradigmatic case of Whig historiography of science. It seems to illustrate vividly how a naive and antiquated myth can be overthrown by the persistent application of the experimental method. The story and its moral have become classics. The historians told how the Aristotelian belief in spontaneous generation was gradually eroded through a series of experiments, running from those of Francesco Redi in the seventeenth century, through Lazzaro Spallanzani in the eighteenth, to the climactic figure of Louis Pasteur in the nineteenth century. In addition, since Redi, Spallanzani, and Pasteur experimented on insects, infusorians, and bacteria respectively, the necessary progressive aspect of science history became apparent. From Redi to Pasteur it was thought that smaller and smaller organisms were capable of being generated spontaneously. That few other episodes in the history of biology illustrate so vividly the moral of progress through the application of experiment probably explains why so many introductory biology texts still find room to describe how Pasteur and his precursors overcame the ancient myth. As one such text recounts the events:

It was only in the last century or two that serious doubts arose about the origin of life. The practical man knew that new life appeared around him all the time: flies and maggots from rotting meat and barnyard manure, life from sweat, glow worms from rotting logs, eels and fish from sea mud, and frogs and mice from moist earth. No less an authority than Aristotle vouched for such commonsense observations, and for over two thousand years, SPON-TANEOUS GENERATION was accepted as a fact of nature.

A few people questioned whether spontaneous generation ever really occurred. Francesco Redi, a Tuscan doctor, demonstrated in 1668 that maggots in meat were only the larvae of flies, and that if the meat was protected, so adult flies could not lay their eggs, no maggots appeared. When the Dutch lens grinder and microscope maker Antoni van Leeuwenhoek discovered microorganisms in 1676, spontaneous generation received new support. While many people were ready to concede that horses and maggots did not appear spontaneously from nonliving matter, they were less confident about the subvisible creatures that van Leeuwenhoek could find everywhere. The Italian biologist Lazzaro Spallanzani (1729–1799) showed that if broths were adequately sterilized and kept from being contaminated by air-borne microorganisms, they remained devoid of life. But he failed to convince his contemporaries, partly because others were performing the same experiments with less care and obtaining exactly the opposite results. Experimental methods were not good enough to rule out spontaneous generation for those who wanted to believe in it, as many people did for philosophical reasons.

Louis Pasteur finally laid spontaneous generation to rest permanently in 1862, in response to competition for a prize set up by the French Academy of Science. Pasteur won the prize for a series of meticulous and conclusive experiments that showed that microorganisms came only from other microorganisms, and that a genuinely sterile broth or solution would remain sterile indefinitely unless contaminated by living creatures. The old aphorism, *Omne vivum ex vivo* (All life from life), became dogma.[3]

Unhappily the historians who tell this story ignore the warning of Claude Bernard. "The experimenter," he wrote, "who does not know what he is looking for will never understand what he finds."

The classical story has never presented the concept of spontaneous generation very seriously.[4] There seems to have been a tacit understanding that had more scientists been prepared to observe nature, then the issue would never have arisen and would certainly not have survived so long. Indeed, one well-known scientist writing a few years ago on Georges-Louis Leclerc de Buffon, an eighteenth-century supporter of spontaneous generation, declared:

Buffon was contemptuous of the study of these lowly organisms. "The discoveries that one can make with the microscope," he wrote, "amount to very little, for one sees with the mind's eye and without the microscope the real existence of all these little beings...." It was fortunate for the progress of biology that there were others who did not share this opinion.... They examined not with the mind's eye, but with the microscope. By observing... [they] struck a powerful blow against the theory of spontaneous generation.[5]

The issue of spontaneous generation was not to be denied in that manner, however. Indeed, mere observation was more likely to lend support to the doctrine than to deny it. In 1832, the German physiologist, Carl Burdach, rightly remarked that "those who defend the doctrine of spontaneous generation do so through experience."[6]

The demise of the concept of spontaneous generation cannot be ascribed merely to the collection of experimental and observational data. Indeed, the relationship between spontaneous generation and experimentation is a complex one. Opponents of spontaneous generation, by arguing that all organisms arise from parents or, conversely, that no organism arises spontaneously from matter, were faced with a logical dilemma in that both these statements can be falsified but neither of them can be proven with absolute certainty. On the other hand, their opponents, by arguing that some organisms can indeed arise directly from matter, held to a belief that cannot be falsified but can be proven. Logically speaking, therefore, opponents of spontaneous generation could do no more than invalidate particular experiments said to illustrate its occurrence. Paradoxically, although the logical problems are widely recognized, standard accounts of the controversy persist in claiming that the belief in spontaneous generation was overthrown by experimental proof.

The questionable significance of experimentation is particularly obvious in arguments over the origin of internal parasitic worms, a group of organisms long regarded as displaying the validity of spontaneous generation. In this case no experimental tradition was established prior to 1851, for the very good reason that neither side of the controversy could hope to achieve much by subjecting the issue to experimentation. Opponents of spontaneous generation, wishing to show that worms were derived from material taken in with air, food, or water, could prove nothing by feeding a potential host animal with such material. So long as an origin by spontaneous generation was a feasible alternative, there was no way to prove that the animal was not infected prior to the experiment, and no way to prove that any worms found subsequently were in fact derived from the ingested material. Conversely, proponents of spontaneous generation, wishing to deny that worms were derived from external sources, could not prove this without preventing access to air, food, and water. Even suggestive experiments, such as feeding one group of organisms with living worm eggs and another group with killed worm eggs, were rarely, if ever, carried out. Such experiments were doomed to failure in any case, in part because internal worms are not passed directly from host to host but go through intermediate stages in other organisms, and in part because proponents of spontaneous generation never denied that parasites could sometimes be passed from host to host.

The discovery of heat-resistant bacterial spores in the 1870s has been seen traditionally as the final conclusive vindication of Pasteur's beliefs. Their existence seemed to explain many of the anomalous results obtained by Felix Pouchet and Henry C. Bastian. Yet the determination of thermal "death points" was once again beyond the scope of experimentation. These crucial experiments rested on one essential question: "What is the lowest necessary amount of heat to ensure, on the one hand, that all pre-existing living things shall be killed and to avoid as far as possible extreme deterioration by heat of the fluids employed?"[7] Proponents of spontaneous generation argued that heating destroyed the ability of fluids to engender life and, thus, that boiling for long periods would render such fluids sterile. But, if boiling was limited to a short period, such fluids could and often did engender life. Opponents of spontaneous generation, however, argued that any appearance of life after boiling was a result of the survival of such organisms or their spores in the boiling medium. Both sides of the debate, therefore, drew conclusions from the experiments by begging the very question at issue. Those arguing for survival of spores did so on a prior rejection of the possibility of spontaneous generation, and, conversely, those arguing for the "germinality of fluids" did so from an a priori acceptance of spontaneous generation. Any determination of death points, then and now, is only possible if the possibility of spontaneous generation is denied. Once again the issue in question remained insoluble by experimental means.

As a result of these problems the experimental evidence against the occurrence of spontaneous generation has had different impacts at different times. Generally, however, scientists have been more willing to accept the experimental evidence against spontaneous generation than is logically entailed in these results. If we accept that a scientist's hypotheses or theories must explain data, "in terms of which his data will fit intelligibly alongside better known data,"[8] the reasons for this become clear. Opponents of spontaneous generation always fulfilled the required goal, since they could argue from analogy that *all* living things arise from parents. On the other hand, those supporting the doctrine were always forced to justify an anomaly. They could claim only that *some* but not the majority of organisms arise without parents. Clearly this was an inherent weakness in their argument, which continually undermined their case and gave credence to their opponents' experiments. These experiments also carried favor when arguments from other scientific and nonscientific areas seemed to substantiate the impossibility of spontaneous generation. Thus, in the 1870s, for example, these experiments loomed large in Britain and Germany as the complex nature of the simplest cellular forms became realized. The French in the previous decade were ready to accept the validity of Pasteur's claims partly through their adherence to orthodox religious and political views.

However, when the occurrence of spontaneous generation seemed a distinct possibility, scientists were more ready to point out the logical impossibility of disproving the belief in spontaneous generation. This occurred either when some particular metaphysical position came into vogue (nineteenth-century materialism and *Naturphilosophie*), or when the anomalous few that could arise spontaneously were delimited from the other organisms by their simplicity of form, only a step in complexity above nonliving material.

Despite these problems, the classical story has dwelt almost exclusively on experimental details of the debate, with very little analysis of why the experimenters believed or disbelieved in the phenomenon in the first place. Failing to stress that experiments are designed to validate a previously held position, they omitted to mention, for example, that Spallanzani experimentally attacked spontaneous generation from an a priori belief in the doctrine of preexistence. In addition, the classical story has ignored almost entirely the numerous nonexperimental issues that were fundamental to the general debate.

This silence results in part from the actions of Pasteur himself, since the classical story is derived from his historical introduction to the justly famous paper of 1861, "Mémoire sur les corpuscules organisés qui existent dans l'atmosphère." As discussed in chapter six, this strangely incomplete account may well have reflected a conscious or unconscious desire on the part of Pasteur to steer the argument away from certain critical areas and toward those experimental issues with which he was concerned. The prestige of Pasteur has become so great, however, that succeeding generations have accepted his account at face value.

The argument over spontaneous generation incorporated many more issues than those discussed by Pasteur and extended, in time, well past the period in which Pasteur worked. Neither is the doctrine one that has been slowly eroded since the heyday of Aristotelianism. Indeed, the doctrine enjoyed its widest popularity in the early years of the nineteenth century and was decidedly unpopular in the preceding years. It also received substantial support in the 1860s and again in the 1920s. This extension of the debate into modern times reflects the impact of the theory of evolution on questions concerning the origin of life. The very year that Pasteur began his attacks on the doctrine, Charles Darwin's *Origin of Species* appeared. This seemed to some to entail the assumption that life must have originated by a spontaneous generation. Before Darwin, the controversy was over the origin of present-day organisms, although even then it became associated with Lamarckian evolution. After Darwin, the controversy also involved the origin of life. That Darwinism, with its implication of a spontaneous origin of life, appeared at the very time when many felt the doctrine of spontaneous generation had been destroyed, introduced mass confusion into the debate.

It is true that, after 1880, the belief that complex cellular organisms could arise spontaneously never again appeared, and in that very restricted sense the issue of spontaneous generation was abandoned. However, in the early years of the twentieth century, work on colloids and viruses suggested that life may exist at a subcellular level and that subcellular living entities had arisen by spontaneous generation in the past and might well arise in the present. Although some were reluctant to label such a belief by the term *spontaneous generation,* preferring to restrict that term to its nineteenth-century meaning, it is clear that both beliefs are essentially similar. Proponents of both views believed that a fully-formed living entity, having all the attributes of life, arose all at once from matter lacking any such attributes. There are those today who persist in this belief although only in reference to the origin of the first living entity. They claim that in the past a "living molecule," endowed with the characteristics of life, arose by a fortuitous combination of molecules. Thus, contrary to popular belief, spontaneous generation still has its supporters among modern scientists.

Most scientists today have abandoned such a belief. They have accepted the evolutionary concept as expressed by the Russian Marxist biochemist Aleksandr Oparin. They deny that life arose all at once by an improbable but not impossible combination of molecules. Instead, they believe that life slowly emerged as a necessary result of a long-term chemical evolution. In the words of the English physiologist J. W. S. Pringle: "Life truly evolves from chaos with no moment of special creation to identify its birth."[9]

That spontaneous generation is no longer an issue to most biologists, stems, in the final analysis, from the work of Oparin, not Pasteur. It is of interest that, in the seventeenth, eighteenth, and nineteenth centuries, the controversy involved deep religious issues, but that, in most recent times, political issues played a greater part. This is not surprising, for arguments over spontaneous generation have always incorporated issues beyond the confines of sterile flasks; the concept has called into question some of mankind's most cherished beliefs.

Donnez-moi de la matière et du mouvement,
et je ferai un monde. VOLTAIRE

TWO

ABHORRENCE OF CHANCE

THE SEVENTEENTH AND EIGHTEENTH
CENTURIES

I n an age dominated by common-sense observations and interpretations, there was no need to question the occurrence of spontaneous generation. It was an observational fact. Eels were generated out of mud, fleas and lice appeared on the bodies of man and beast, and parasitic worms likewise arose internally. Insects sprang from plant galls and decaying plant and animal matter, fungi from trees and earth. Pond water generated an awesome variety of animal and plant life. Such beliefs were commonplace at the beginning of the seventeenth century, at which time explanations of such events derived from antiquity. But by the end of the century *all* explanations of spontaneous generation had become incompatible with the dominant new scientific philosophy, a philosophy based on a deistic mechanism. Thus, despite continued empirical evidence in its favor, belief in spontaneous generation collapsed. By 1704 it could be reported, "The learned world begins now to be satisfied, that there is nothing like this in nature."[1]

In the early years of the seventeenth century however, spontaneous generation seemed to provide a particularly clear case in support of the new mechanical philosophy. René Descartes, who reduced natural science to the consideration of matter and motion alone and ruled out all "occult" and final causes, felt that for spontaneous generation to occur it was merely necessary for heat, acting on putrefying matter, to agitate the subtle and more dense particles to form an organic being. Similarly, sexual generation and embryological development were seen in terms of matter and motion: "If one knows well what are all the parts of the seed in a certain species of animal, for example of man, one could deduce from that alone, by reasons entirely mathematical and certain, the whole figure and conformation of each of its members."[2]

Acceptance of spontaneous generation was also an important aspect of Paracelsian teaching, which enjoyed a wide vogue in the seventeenth century. With generation of any natural object being part of a

8

universal chemical putrefaction, there was nothing particularly unique or unexpected about organisms generated spontaneously. Just as milk was the matrix or "mysterium" out of which cheese was generated, so cheese was the "mysterium" of maggots and worms. "All in turn descend from the 'Mysterium Magnum' which is the one mother of all things and of all elements and a grandmother of all stars, trees and creatures of the flesh."[3] While it was commonly believed in antiquity that, "after a long period of fertility, during which many monstrous and marvellous generations were brought forth, the Earth mother became at last exhausted and sterile and lost her power of producing men and the larger animals,"[4] retaining only enough vigor to produce the small organisms, no such limitations were placed on its operations by the Paracelsians. By viewing spontaneous generation in terms of putrefaction, it became universal and unlimited. Even an artificial man, or *homunculus,* could be prepared in a laboratory vessel from putrefying male semen sustained by the "arcanum" of human blood.

Paracelsian and other iatrochemical views, like those of the mechanists, were part of the new revolt against overreliance on the ancients and in favor of more emphasis on observation and experiment. In France and England much of the Paracelsians' spleen was directed at the physicians with their reliance on Galenic teachings. In the 1640s and 1650s a large number of iatrochemical tracts appeared in England, included in which were the posthumous works of Jean-Baptiste van Helmont.[5]

As P. M. Rattansi has pointed out, much of the popularity of Paracelsian and Helmontian teaching in England can be explained in terms of the conflict between the physicians and apothecaries of London. Iatrochemical doctrines provided a rationale for the therapy of the apothecaries and gave extra weight to their attacks on the practice and teachings of the Galenic physicians. Iatrochemical teachings also struck a receptive chord among the Puritans and provided a mystical and religious alternative to the mechanical philosophy.[6]

Robert Boyle himself was impressed by the theories and experimentalism of van Helmont, in particular the famous willow tree experiment.[7] In this experiment, a willow shoot weighing 5 pounds was placed in 200 pounds of soil and allowed to grow. After five years, during which time only water was added, the weight of the soil was virtually unchanged, yet the willow shoot had increased its weight to 168 pounds. "All vegetables," van Helmont concluded, "do materially arise wholly out of the element of water."[8]

Thus, in the early years of the seventeenth century, there was no reason to question the occurrence of spontaneous generation. It was explicable under both the older Aristotelian and Galenic philosophies and the more modern concepts of Cartesian mechanism and iatrochemistry. Such a happy turn of events was, however, short lived. By the middle of the century, not only were the explanations of antiquity becoming more and more suspect and the iatrochemical viewpoints on the wane, but also mechanical concepts seemed inadequate to explain the occurrence of spontaneous generation. There was no longer room for Aristotelian "souls," Galenic "faculties," and "indwelling intelligencies," by which the ancients had explained animal generation in general and spontaneous generation in particular. For example, despite the general fruitfulness of his "primordium" concept, William Harvey's general views on animal generation had very limited success. He was seen as an Aristotelian attempting to perpetuate an outmoded doctrine. As Jacques Roger remarks, Harvey was "étouffé en naissant par l'esprit mécaniste du siècle."[9]

During the same period the popularity of iatrochemical teachings in England declined rapidly. After the Great Plague of 1665, it played an ever-decreasing role in late seventeenth century thought.[10] Even more significant, the continuing strife between the increasingly powerful apothecaries and the College of Physicians led the latter to switch their allegiances from traditional Galenic theories to the more "modern and progressive" mechanistic concepts.[11] As such they joined forces with the Royal Society, where the corpuscular theories of Robert Boyle and Isaac Newton dominated.

Similarly in France, the new center of scientific learning at l'Académie Royale des Sciences, like its English counterpart, became an advocate of the "new philosophy." Mechanism was "a constant maxim among modern philosophers."[12]

Few of these modern philosophers could accept Descartes' explanation of generation or spontaneous generation, however, even those who considered themselves Cartesians. It seemed inconceivable that the complex form of a living being could ever be produced solely by the general laws of motion. As George Garden remarked in 1691: "All the laws of motion which are as yet discovered, can give but a lame account of the forming of a plant or animal. We see how wretchedly Descartes came off."[13] Or again, in 1683, an anonymous writer in the Philosophical Transactions of the Royal Society stressed that, since "every animal ... is made up of so many different parts, and those of so excellent a contrivance, and so wonderful a respect to one another," it would be inconceivable "that the seminal fluids lying aloof and at large in the capacity of the womb, and exposed to so many accidents could give a production so admirable."[14]

This emphasis on complexity reflected the impact of the new

anatomical and microscopic findings on seventeenth-century thought. The widespread interest in insect anatomy and the new world revealed to naturalists by the microscope "displayed such a complexity of organization in the lowest and minutest forms, and everywhere revealed such a prodigality of provision for their multiplication by germs of one sort or another, that the doctrine of [spontaneous generation] began to appear not only untrue, but absurd."[15]

Thus, as we move into the second half of the seventeenth century, the problem of spontaneous generation, and more importantly sexual generation, became acute. In neither case did Cartesian explanations seem feasible while nonmechanistic explanations were no longer acceptable. It was the ensuing debate over sexual generation that eventually destroyed the belief in spontaneous generation. This debate reflected the impact of "un nouvel esprit scientifique" that took root in the late seventeenth century.[16]

The new spirit was based primarily on the Cartesian belief that matter was purely passive; that all phenomena resulted from the motion of this passive material. In addition, all laws of motion were considered immutable and uniform in all of nature. As Antonio Vallisneri remarked, "Nature in all things is one, pure, simple, and immutable."[17] Spontaneous generation seemed to be at odds with this concept of nature. That organisms could arise spontaneously "by chance" implied that such occurrences were without cause. They were accidental, exceptional, unlawful, unknowable events, which ran contrary to the widely held concept of the universality of natural law. In addition such organisms displayed the usual specificity of type and were not a random hodgepodge of individuals. How, then, could unlawful and accidental events repeatedly lead to the production of reoccurring and limited types?

Thus the belief was established that generation is the same in all animals, that species are fixed, and that one could admit of no irregularities in nature. In addition, a strong observational and experimental tradition grew up which seemed to substantiate these new a priori concepts. The result was that "experience and reason together condemned all spontaneous generation as contrary to the infallible regularity of nature."[18] It led eventually to the astonishing success of the doctrine of germ preexistence, and thus to the denial of any form of spontaneous generation.

The term *preexistence* has long been confused with *preformation*. Both were a response to the problem of organic form which presented such a serious problem to the new mechanistic thinking. Preformation teaches that development is the mere mechanical growth of a miniature preformed *in the parent organism*. It first appeared in the early seventeenth century and was the doctrine to which Harvey was opposed and which Kenelm Digby attacked in 1644. Indeed, it is interesting to note that the latter regarded the occurrences of spontaneous generation—vermin in living bodies, rats in ships, frogs in air, eels in mud—as the most obvious refutation of preformationist concepts.[19]

Preexistence, however, flourished in the late seventeenth century and had little in common with early preformationist concepts.[20] Preexistence teaches that the germ of preformed parts is not produced by the parent, rather it is created by God at the beginning and is conserved in that state until the moment of its development or *l'évolution*.

Preexistence theories first appeared as an attempt to explain so-called spontaneous generation, and only after the appearance of ovist and spermist preformationist ideas did it become the explanation of all generation. Both Daniel Sennert and Pierre Gassendi held to such preexistent theories in their dealings with spontaneous generation. The former held to preexistent souls, the latter to preexisting germs. On the one hand, Sennert, an early atomist, believed the atoms alone were incapable of arranging themselves to form a living being; they must be animated by a soul. Since a soul cannot be produced from matter, only from God, and since God had ceased to create, there must be an uninterrupted chain of souls running back to the first creation of each species. Thus spontaneous generation can only occur in matter endowed with a soul, that is, in living matter or matter that was once alive.

Gassendi, on the other hand, refuted the concept of spontaneous generation. Denying that God created only atoms at the beginning, he argued instead that God created seeds, from which "we can certainly understand how it comes about that the animula [soul] contained in the seed begins, prosecutes and perfects with such great skill and industry the elaboration of its organs and of its whole body." Thus he denied a true spontaneous generation. "The animals of both kinds owe their origin to their own seeds, but those of the first [normal generation] are customarily said to arise from seeds, those of the second [spontaneous generation] from putrid or other material, since the seed of the first is visible within a similar animal, whereas the seed of the second is hidden in foreign and unexpected material."[21]

But it was the concept of the universality of ova, stemming from the suggestion of William Harvey, that led eventually to late-seventeenth-century preexistence theories and the necessary denial of spontaneous generation. As Elizabeth Gasking and Everett Mendelsohn so rightly remark, Harvey's *Generation of Animals* was not directed toward the problem of spontaneous generation. His concern lay with sexually reproducing forms. There are numerous passages in the book that make reference to spontaneous generation, but while some seem to deny it others equally seem to permit it.[22]

Galenists and Aristotelians were in agreement that the offspring of sexually reproducing forms developed from an intermingling of male semen and menstrual fluid. Whether the material of the off-spring was derived from the mixing of the two semina or whether the male semen played a nonmaterial role was the area of disagreement. Harvey, however, discovered by dissection that no mass of semen could be found either in the chick or deer after intercourse. From this he concluded that any theory of generation could no longer involve the totality of male semen and menstrual fluid. In addition he found no changes in the "female testes" (i.e., ovaries) of the deer and concluded that they played no role in generation.

Harvey subsequently directed his main thrust against the belief that there were three modes of animal reproduction: vermiparous, oviparous, and viviparous. Rather, he argued, all animals are a prod-uct of an egg or primordium. That, while an egg is "a conception exposed beyond the body of the parent, whence the embryo is pro-duced; a conception is an egg remaining within the body of the parent until the foetus has acquired the requisite perfection." Thus "all em-bryos are procreated from some conception or primordium."[23]

The primordium was "a certain corporeal substance, from which, through the motions and efficacy of an internal principle, a plant or animal is produced."[24] The role of the semen was seen in terms of its capacity to lead to a complete development; the primordium itself, contrary to Aristotle, containing both the necessary matter and the efficient cause. Like many of the mechanistics themselves, to whom Harvey was violently opposed, he dismissed Cartesian-like expla-nations as absurd. "There is a greater and more divine mystery in the generation of animals, than the simple collecting together, alteration, and composition of a whole out of parts would seem to imply."[25]

Following the work of Harvey there seemed to have been general agreement among anatomists over the universality of "eggs," thereby rendering spontaneous generation an anomaly. It is also clear that the "female testes" were often considered to be the source of these eggs and not the uterus as Harvey had suggested. Both Thomas Wharton in 1656 and Thomas Kerckring in 1671 suggested that semen joins with the egg in the "female testes" after which it descends to the uterus.[26] By 1670 the presence of vesicles in the female testes was widely recognized, but it remained obscure whether "these vesicles fastened to the body of females . . . wherein the embryo is formed" were some of "the vesicles loosened [from the female testis]."[27] In

1672 Régnier de Graaf demonstrated to many peoples' satisfaction that the attached vesicles were derived from the "female testes" and, thus, that these so-called testes were ovaries.[28]

Acceptance of ovism was rapid, so that by 1679 it could be stated that "the view of the formation of man, as well as all the other animals by means of eggs, is so common at present that there are almost no philosophers who do not admit it today."[29] Ovism, coupled with the strong a priori belief in the uniformity of natural law that would admit no exceptions, irregularities, or chance occurrences, led necessarily to a denial of spontaneous generation.

Such denials received strong empirical support at this time from the work of Francesco Redi, Jan Swammerdam, and Marcello Malpighi. The physician Redi, born at a time when the Jesuits were engaged in their polemic with Galileo and subsequently trained by the Jesuits, may have been led into the spontaneous generation controversy through his abhorrence of extreme Paracelsian teachings. His initial experimental attack on the doctrine rested on the belief that:

the earth, after having brought forth the first plants and animals at the beginning by order of the Supreme and Omnipotent Creator, has never since produced any kinds of plants or animals, ... and everything ... that she has produced, came solely from the true seeds of the plants and animals themselves.... And although it be a matter of daily observation that infinite numbers of worms are produced in dead bodies and decayed plants, I feel, I say, inclined to believe that these worms are all generated by insemination.[30]

Placing an awesome variety of dead animal material in open boxes—snakes, pigeons, fish, sheeps' hearts, frogs, and meat from deer, dogs, lambs, rabbits, goats, ducks, geese, hens, swallows, and even lions, tigers, and buffalo, he was struck in every case by the occurrence of the same kind of flies hatching from maggots engulfing the rotting flesh. Noticing also the general presence of eggs on the flesh, he remarked: "These eggs made me think of those deposits dropped by flies on meats, that eventually became worms, a fact ... well known to hunters and to butchers, who protect their meats in summer from filth by covering them with white cloths."[31] He was able to verify this hypothesis by preventing the appearance of maggots on meat in closed and sealed vessels and in vessels covered by a net. In the latter case maggots appeared on the outside of the net where he also saw insects depositing their eggs and maggots. Thus he concluded that spontaneous generation of fly maggots from dead flesh did not occur, but rather, "All dead flesh, fish, plants and fruits form a good breeding place for flies and other winged animals."[32]

On the other hand, he was not prepared to deny the occurrence of animals generated from living flesh. He likened the production of gall flies in plants to "the peculiar potency of that soul or principle which creates the flowers and fruits of living plants." Redi reasoned: "If a thing is alive, it may produce a worm or so, as in the case of cherries, pears and plums; in oak glands, in galls and welts of osiers and ilexes worms arise, which are transformed into butterflies, flies and similar winged animals. In this matter I am inclined to believe, tapeworms and other worms arise, which are found in the intestines and other parts of the human body."[33]

The work of Swammerdam and Malpighi quickly removed gall flies as examples of spontaneously generated organisms. Swammerdam "believed absolutely" that:

it is not possible to prove by experience that insects are engendered out of plants. But on the contrary we are certain that these small animals are enclosed there only to take nourishment and there is even the probability that these same plants are created only for this purpose. It is even true that by a constant and immutable order of nature we see regularly several sorts of insects attached to specific plants and fruits . . . they come from the seed of the

A plant gall opened to display the insect larva trapped within. Francesco Redi and others regarded such insects as a product of spontaneous generation. Courtesy M. J. Harvey.

animals of their own species which were laid there earlier. These insects insert their seed or eggs so far into the plant, that they consequently unite with them and the aperture closes. The eggs nourish themselves within.[34]

Swammerdam, and all other naturalists working on insects, constantly reiterated their opposition to chance occurrences in nature. Struck by the pains that insects took to deposit their eggs in places suitable for their development, Swammerdam remarked "that no generation happens in nature by chance, but by propagation and by growth of parts where chance has not the least part."[35]

Ovist and experimental opposition to spontaneous generation gained additional and strong support at this time with the appearance of numerous tracts on natural theology. John Ray's *Wisdom of God manifested in the works of creation* appeared in 1691, and in 1713 his *Three Physico-Theological Discourses* appeared. In the same year the first edition of William Derham's *Physico-Theology or, a demonstration of the Being and Attributes of God from his works of Creation* was published.

These texts appeared at the same time as the preexistence theories came into vogue. To both natural theologians and supporters of preexistence theories the problem of animal form was not resolvable by mechanistic means, but only by reverting to the "infinite and varied expressions of divine foresight."[36] John Ray, for example, attacked the Cartesians or "Mechanic Theists," remarking with justification: "But the greatest of all the particular phenomena is the formation and organization of the bodies of animals, consisting of such variety and curiosity; that these mechanic philosophers being no way able to give an account thereof from the necessary motion of matter, unguided by minds for ends, prudently therefore break off their system there."[37]

Once the origin of animal form reverted to God himself, then clearly any spontaneous generation became an act of creation. Since creation was "the work of omnipotency . . . it must be beyond the power of nature or natural agents," and since creation stopped on the seventh day, then a spontaneous generation became inconceivable on theological as well as scientific grounds.[38] Not surprisingly, therefore, John Ray mounted a massive attack on the doctrine in the second edition of his *Wisdom of Creation*. At this time he spoke of animal generation being under the influence of a "plastic nature," but by 1713 he came to a lukewarm acceptance of ovist preexistence.

The birth of preexistence or *emboîtement* theories was almost inevitable given the absurdity of Cartesian explanations and the lack of explanation presented by the preformists. The latter group was still faced with the problem of explaining how the parent organism could give rise to the preformed embryo. Much confusion still exists as to the originator of the preexistence concept although, as pointed out above, both Sennert and Gassendi had been forced earlier to utilize preexistence-like explanations in their dealings with the problem of spontaneous generation. There seems little doubt, however, that the concept was given its most precise formulation by Nicolas Malebranche in 1673, who, starting from a typically Cartesian position, stated that:

Our sight is extremely short and limited, but it ought not to prescribe limits to its object. The idea it gives us of extension has very narrow bounds, but it does not from thence follow the bounds of extension are so. It is doubtless infinite in a certain sense, and that diminutive part of matter which is hidden from our eyes is capable of containing a world in which may be hid as many things . . . as appear in this great world in which we live . . . and perhaps there may be in nature less and less still to infinity. . . . We have evident and mathematical demonstration of the divisibility of matter *in infinitum,* and that is enough to persuade us there may be animals, still less and less than others *in infinitum.*

He extended his thoughts into the problem of generation:

We may with some sort of certainty affirm, that all trees lie in miniature in the cicatride of their seed. Nor does it seem less reasonable to think that there are infinite trees concealed in a single cicatride since it not only contains the future tree whereof it is the seed, but also abundance of other seeds, which may all include in them new trees still, and new seeds of trees . . . and thus *in infinitum. . . .* We ought to think, that all the bodies of men and of beasts, which shall be born or produced till the end of the world, were possibly created from the beginning of it. I would say, that the females of the original creatures were, for ought we know, created together, with all these of the same species which have been, or shall be, begotten or procreated whilst the world stands.[39]

According to Roger, after 1673 the doctrines of ovism, ovist preformation, and ovist preexistence became thoroughly confused one with another. Peter Bowler points out that even those holding to the doctrine of metamorphosis, such as Malpighi, came to be seen as supporters of preexistence theories.[40]

The popularity of ovism and the decay of twin-semen theories made very obscure the role of the "frothy white liquid" emitted by the male. Indeed the disavowal of a material role for the semen brought back into vogue Aristotelian-like concepts. The discovery of "animalcules" within the semen came as a complete surprise, but their presence did not imply any role in generation.

Although discovered by Antoni van Leeuwenhoek in 1677, the

first public awareness of animacules came from the pen of Christiaan Huygens, to whom Nicolas Hartsoeker had written in March 1678 describing his discovery of sperm.[41] Huygens made no mention of Hartsoeker in this paper, but a little later a second paper appeared in which Hartsoeker was given the credit for the discovery.[42] The appearance of these papers may well have prompted Nehemiah Grew, the secretary of the Royal Society, eventually to publish Leeuwenhoek's original letter two years after it was received.[43]

By March 1678, Leeuwenhoek had decided that "it is exclusively the male semen that forms the foetus," and the eggs "to have no other purpose than to serve as food for the semen of the cock."[44] The origin of the animalcules, however, was a mystery to Leeuwenhoek, although it was discussed at length in his letter to Robert Hooke in 1680.[45]

Belief in animalculism was slow to grow and was never widespread, despite its obvious appeal to the male ego. In an age dominated by preexistence theories and natural theology this was not surprising, since male emboîtement implied an enormous wastage seemingly contrary to a nature "which doth nothing without a purpose" and which abhors chance. That sperm were not always seen added credence to the idea that they were accidental harmful parasites of the testes. But animalculism did have a role for the egg—a nonpreformed egg—in that the egg proved a suitable nidus in which the sperm could develop. Perhaps this aspect of compromise was the strongest argument in its favor. Certainly none other than the Lexicon Technicum favored male emboîtement and repeated verbatim the paper by Garden which appeared in 1691.[46] Surprisingly, too, William Derham supported animalculism in his text on natural theology.

Perhaps the most significant animalculist, besides Leeuwenhoek and Hartsoeker, was the physician Nicholas Andry de Boisregard, whose text De la génération des vers appeared in 1700. As a result of this work, he was appointed to the faculty of medicine at the Collège de France.

By the end of the seventeenth century, only the appearance of parasitic worms seemed inexplicable by ovist or spermist means. John Ray remarked in 1691 that, although "there is no such thing in nature as aequivocal or spontaneous generation," nevertheless, "no instance against this opinion doth so much puzzle me, as worms bred in the intestines of man and other animals."[47]

The life cycle of the parasitic flukes (trematodes) and of tapeworms (cestodes) involves passage through one or more intermediate hosts and drastic changes in body form. (See appendixes.) Before the discovery of these facts in the middle of the nineteenth cen-

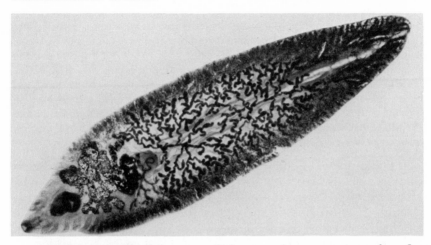

An adult parasitic fluke, one of the most important examples of a spontaneously generated organism.

Nothing except death can prove with a greater strength of evidence, that he is fallen from his original state of integrity and favour with God, than such an army of scourges set in array against him.—w. KIRBY, *The Bridgewater Treatise* 1835

tury, there was no rational way to explain their origin in man and other animals except by spontaneous generation. In an age that rejected any such possibility, parasitic worms presented an enormous problem. As late as 1843 Richard Owen remarked that "the hypothesis of equivocal generation has been deemed to apply more strongly to the appearance of intestinal parasites in animal bodies than to the origin of animalcules in infusions."[48] The appearance of Andry de Boisregard's text was therefore of considerable significance. Not surprisingly Andry could mount little substantive evidence to support his statement that "worms breed in the bodies of men and other animals, by means of a seed that enters there, in which these worms are enclosed."[49] The text was more concerned with "the Nature, Several Sorts, Effects, Symptoms and Prognostics" of the worms; the question of their origins was discussed with reference to animalculist preexistence. Indeed, the book perhaps is seen better as a defense of this doctrine than as one devoted to the worm question per se.

The essential assumptions behind Andry's rejection of spontaneous generation were best expressed by the iatromechanist Giorgio Baglivi, whose letter to Andry was appended to the text:

It is a shame to philosophers and physicians, that in this most happy age of sciences, wherein the causes of things are illustrated by experiments and solid precepts of mathematics, to ascribe to the fortuitous change of putrefaction, that which the constant and perpetual laws of nature remaining in its seeds, rules and directs. For it is not putrefaction that produces imperfect living creatures, but the heat and fermentation of things putrefying that makes fruitful the seeds of things spread all over the world. . . . What we have said before of insects in general, may be rightly applied to the worms bred in human bodies, seeing they are not generated by putrefying humors, as the Pseudo-Galenists commonly think; but the worm eggs lying hid in the intestines, are enlivened and brought forth by the same means.[50]

The problem of chance and wastage implied in ubiquitous seeds finding root only in certain specific host organisms did not seem to be such a major problem to animalculists, since the sperm had exactly the same problem in finding the egg in which to develop. Indeed, the parallel between a parasitic host and an egg was often made, each providing a suitable *nidus* for the parasite and sperm to develop respectively. Furthermore, to an animalculist the sperm were parasites, howbeit useful ones; Andry himself divided the worms into two groups, flesh eaters and "spermatiks." The common assumption of historians that sperm were seen as generative beings *or* parasites is not correct; the real distinction was whether the sperm were necessary, useful parasites or accidental, harmful ones.

The problem of wastage that accrues if the worm seeds enter with food and water—or even through the skin as Andry believed—could, however, be solved by another mode of entry. The worm seeds could be assumed to pass directly from host to offspring during copulation, lactation, or across the placental barrier. "Why might not one say, that in case the seed of the worm did not enter the patient's body along with the vituals, perhaps it might have accompanied the blood of his father from the time of his conception. . . . May we not rather conceive that these very same seeds were created in the seed of man, along with man himself."[51]

Such a method of transmission tied in beautifully with ideas on preexistence. In a kind of "double *emboîtement*" theory the worms "are created with man, and their species is perhaps as ancient as mankind."[52] This theory, in which poor Adam not only contained all of mankind-to-be but also all his worms-to-be, on the one hand, removed the problem of chance and wastage but, on the other, presented an awful theological dilemma:

It is not reasonable to suppose that God would have placed the first worm in [Adam's] body, forasmuch as Man in this state of innocence was to be free of all kinds of diseases, in a perfect state of happiness. . . . But if on the other hand, after the lapsed state of Adam, we allow that worms were formed by God . . . a greater difficulty will arise; for hence it will follow, that God made a new creation of worms, which is contrary to Holy Writ; since God hath taught us, that before man was made, all other animals, were created. Therefore worms could not be created in the body of Adam; either before or after his Fall.[53]

Vallisneri attempted to extricate the theory from this dilemma by assuming that the worms did good works before the Fall.

Adam could support and feed those insects which had a mind to live together quietly and friendly, as we may say; and if anything superfluous remained, that they might eat . . . and not transgress their bounds or eat holes through sides of the guts . . . but would rather by gently licking the parts and by healing them do their host a kindly office. . . . But this happiness of Adam was but of a short continuance, for disobeying God . . . all things were suddenly changed; so that these worms were made the Ministries of Divine Justice and raised an insurrection upon him . . . so the worms though inhabitants of the human body, had leave given to destroy and become a common enemy to mankind.[54]

This explanation in turn faced considerable difficulties. In the first place, the theory of "worm *emboîtement*" could not be easily reconciled with the fact that not all men were infected by them or with the obvious differing geographical distribution of different parasites. This suggested to Daniel LeClerc that the worm seeds, "received from the first common parent, could lie long hidden in the body of every single man and there, sometimes, as occasion served, be raised up or excluded as there should be need thereof."[55] More significantly, however, Vallisneri's solution to the theological problem posed by "worm *emboîtement*" was unsatisfactory, for Adam before the Fall was in a state of perfect grace and would thereby need no assistance from "gently licking" worms!

One cannot overemphasize the problem posed by these parasitic worms to the orthodox preexistence theories. Their role in the whole spontaneous generation controversy was immense and remained so until the middle of the nineteenth century. This aspect of the debate needs stressing since Whig historians, modeling themselves after Pasteur's historical account of the debate, have continually overlooked them.

Many other discoveries in the early eighteenth century (in addition to the worm question) presented problems to the dominant view. Abraham Trembley's discoveries of budding in hydra and the regenerative powers of some primitive forms were clearly incompatible with preexistence theories, as was the appearance of plant hybrids— the results of crossing the pollen of one species with the female parts of another. Work on plant classification, stimulated by Carl Linnaeus, also resulted in a new awareness of variations in species which was difficult, although not impossible, to reconcile with preexistence theories. On the other hand, preexistence theories gained credence from belief in the "principle of plenitude" as expressed in the concept of the Great Chain of Being,[56] and, in 1740, from the discovery of parthenogenesis in aphids by Charles Bonnet.

The awareness of these anomalies coincided with a new philosophical outlook that began to make itself felt in mid-eighteenth-century France. Much of this stemmed from the increasing popularity of Newtonianism, which resulted in the collapse of Cartesian thought and its fundamental tenet that matter is completely passive. Newton introduced the concept of universal forces acting between corpuscles of matter, a concept the Cartesians found "occult." The implications of the force concept on ideas of generation were considerable, since its presence bestowed on matter a dynamic attribute that was inconceivable to the Cartesian mechanists. It led, for example, to the reappearance of the theory of epigenesis, which had suffered an eclipse since William Harvey's support of it in the middle of the seventeenth century.

Proponents of epigenesis were not only opposed to preexistence theories, but were also in sympathy with the doctrine of spontaneous generation. They believed that germ cells consisted of an homogeneous substance, in which there was no indication of subsequent differentiation, and that the organism developed from this homogeneous mass by a gradual process of progressive growth and differentiation. Thus, as the word "epigenesis" implies, each organism was thought to be a new formation. Clearly, those who believed that an homogeneous substance within an egg could generate a new organism, had little difficulty in extending this to include any organic substance. To supporters of epigenesis, organic matter had the potential to spontaneously generate a living organism. These new and dynamic Newtonian concepts were first formulated in France by Pierre-Louis de Maupertuis and Georges-Louis de Buffon, and by the English Catholic priest John Turberville Needham.

"Why should not a cohesive force, if it exists in nature, have a role in the formation of animal bodies?" wrote Maupertuis in 1745. Opposed to both Cartesian and nonmechanistic interpretations of generation, he discussed the role played by both sexes and an epigenetic embryology. He said nothing about the problem of spontaneous generation, however.[57] It was in the writings of Buffon and Needham that the doctrine of spontaneous generation was once again established. Both of them attacked preexistence theories, noting the numerous anomalies mentioned above, in particular the work of Trembley, and the increasing awareness of the contribution made by both sexes to the offspring. In addition, Buffon rejected "from philosophy all opinion which leads necessarily to the idea of the actual existence of geometric or arithmetic infinity," and explanations that supposed things already had been accomplished. Buffon's view of nature was essentially dynamic, incorporating a constant cycling of "organic molecules," which were continually released from an organism at death and reformed again into new organisms:

There exists in nature an infinity of little organized beings, and that these little organized beings are composed of living organic parts which are common to animals and vegetables ... that organized beings are formed by the grouping of these parts, and that reproduction and generation are therefore nothing but a change of form which comes about simply by the addition of these similar parts.[58]

Form was imprinted on this huge and indestructible mass of organic parts by the *"moule intérieur,"* which Buffon likened to the gravitational force. It was this "mold" that controlled generation, development, and growth. In early life the organic particles were absorbed and employed in augmenting the different parts, the force of the *"moule intérieur"* assuring that each part of the body would receive only those molecules it could use. At the higher levels, therefore, there were different groups of particles representing the different organs. After puberty, when growth stops, the excess particles were passed to reservoirs—the testes and ovaries. Sperms and eggs were not the cause of the new generation. The former were almost "accidental" results of the particles coming together in the testes, the latter were thought also to contain sperm, and their occurrence was described.

When decomposition occurred, the organic molecules, no longer constrained by the mold, found liberty, and later became incorporated into other organisms. This concept provided the basis for the acceptance of a form of spontaneous generation, more properly called heterogenesis. It could occur only by the mingling of organic molecules, matter that was once alive, and not from inorganic brute matter. The free organic molecules in a decomposing body were viewed as agitating the putrefying body and forming by union with brute particles such things as earthworms, mushrooms, and microscopic organisms.

Similar events also could occur in a living organism:

When there are several malfunctions in the organization of the body, which prevent the absorption and assimilation of all organic molecules in the food by the interior mould, these excess molecules, unable to penetrate the interior mould of the animal, reunite with several particles of brute matter in the food and form, as in putrefaction, organized bodies. This is the origin of tapeworms, ascarids, flukes and all the other worms which are born in the liver, stomach and intestines.[59]

Buffon also declared, in another place: "My experiences demonstrate quite clearly that there are no pre-existent germs, and at the same time they prove that the generation of animals and vegetables is not univocal. There are perhaps as many beings produced by the fortuitous mingling of organic molecules as there are [those] which can produce by a constant succession of generation."[60]

Experimental proof of spontaneous generation was provided by Abbe Turberville Needham, whose theoretical explanations of generation rested on "vegetative powers, which reside in all substance, animal or vegetable, and in every part of these substances as far as the smallest microscopical point." Although Needham's explanations involved a vegetative force rather than a combination of organic molecules, both he and Buffon agreed "that there is a real productive force in nature."[61] Similarly, they agreed that the animals found in animal seed were not essential but were rather a "consequence of properties in the seed, which properties were essential to generation."[62] Not surprisingly therefore, Needham, like Buffon, attacked preexistence theories. He argued that both parents contribute to the next generation, and that belief in animalculism implied "so vast an expense, so great a waste of millions of entities." Furthermore, he asked, "if they [animalculae] float in the air, or lie hidden in food, as some have thought, how is it that the Stamina of one species do not sometimes insinuate themselves into a strange parent, with all the inconveniences and absurdities of equivocal generation?"[63] By equivocal generation Needham implied the production of one species directly from the parent of another, a possibility he denied since "the specific semen of one animal can never be molded into another."[64]

This denial of spontaneous generation was restricted to higher plants and animals, however. Microscopical animals on the other hand, "constitute a class apart, and their greatest characteristic is, that they neither are generated . . . or generate in the ordinary way."[65] As a proof of this, Needham

took a quantity of Mutton-Gravy hot from the fire, and shut it up in a phial, closed with a cork so well masticated, that my precautions amounted to as much as if I had sealed my phial hermetically. I thus effectually excluded the exterior air, that it might not be said my moving bodies drew their origin from insects, or eggs floating in the atmosphere. I would not instil any water, lest, without giving it as intense a degree of heat, it might be thought these productions were conveyed through that element. . . . My phial swarmed with life. . . . Let it suffice for the present to take notice, that the phials, closed or not closed, the water previously boiled, or not boiled, the infusions permitted to teem, and then placed upon hot ashes to destroy their productions, or proceeding in the vegetation without intermission, appeared to be so nearly the same, that, after a little time, I neglected every precaution of this kind, as plainly unnecessary.[66]

By distinguishing between univocally produced and equivocally produced organisms, Needham and Buffon attempted to counter the strongest argument put forward by their opponents—that of analogy. The claim, that since higher forms generate only univocally, so must all, now had to be substantiated by experimental proof, a problem of enormous complexity. "The method of reasoning by analogy," Needham remarked, "is but too apt to lead us into mistakes, and therefore we ought to be very diffident of consequences deduced this way."[67]

The experimental findings and theoretical postulates of Needham and Buffon were eventually challenged by Lazzaro Spallanzani. His first experimental refutation of spontaneous generation appeared in 1765, from which beautifully contrived experiments he carefully concluded that:

In hermetically sealed vessels, provided that the enclosed air has not been exposed to heat, one is not always sure of preventing the birth of animals in boiled infusion. If, on the contrary this air has been strongly heated, no animals will be born, at least when no new air is permitted to enter. That is to say air which has not been exposed to heat, is indispensible for the production of animals. And as it will not be easy to prove that there are no small eggs floating in the volume of air contained in the flasks, it seems to me that the existence of these eggs is always suspected and that heat has not destroyed entirely the fear of their existence in infusions.[68]

Four years later a French translation of Spallanzani's work appeared, accompanied by copious and critical notes by Needham.[69] Needham's explanation of the sterility of the infusions rested on a belief that heat destroyed the "vegetative force" in matter and the "elasticity of air."

Spallanzani responded to these criticisms in a long and detailed investigation of spontaneous generation published in 1776. He first investigated "whether according to a new theory of generation, animalcula are produced by a vegetative power in matter."[70] After boiling seven different sets of seed infusions from kidney beans, vetch, buckwheat, barley, maize, marrow, and beets, for a half-hour, an hour, an hour and a half, and two hours, he sealed them loosely with a cork. If boiling gradually destroyed the "vegetative force" in matter, Spallanzani argued, then fewer organisms would appear in those infusions boiled for the longer periods. This did not occur. In many cases infusions boiled for two hours contained more infusorians than those boiled for half an hour. Thus, he concluded, "long boiling of seed infusions does not prevent the production of animalcula."[71] He also found that exposure to the extreme heat of fire did not prevent the subsequent appearance of animalcules. "These facts fully convince me," concluded Spallanzani, "that vegetable seeds never fail to produce animalcula, though exposed to any degree of heat, whence arises a direct conclusion, that the *vegetative power* is nothing but the work of imagination."[72]

In order to discover "whether the influence of heat diminishes the elasticity of air included in vessels hermetically sealed,"[73] Spallanzani drew out the neck of his vessels into fine capillary tubes that could be sealed instantaneously. In that manner the "internal air could suffer no alteration."[74] He then boiled each of nine sets of different infusions for half a minute, one minute, a minute and a half, and two minutes and found that even the half-minute boiling was sufficient to prevent the occurrence of the larger infusorians. Extending these observations he reported that a temperature as low as 167°F. was sufficient to kill these larger forms; while it took as long as three-quarters of an hour of boiling to render the infusion free from all the smaller forms of animalcules. Interestingly enough, this series of experiments led Spallanzani to conclude that the germs of the smaller infusorians could resist the heat of boiling and to conclude, after a further series of experiments, that animals and plants were more easily destroyed by heat than their eggs and seeds.

Spallanzani had to admit, however, that "on breaking the first hermetic seal, a faint hissing was heard," from which he had to suspect "that heat had truly injured the elasticity of the internal air."[75] Although he found the elasticity to have increased by what he ascribed to the release of air contained in the vegetable seeds, it does illustrate the great difficulty of answering all of the objections of Needham.

Spallanzani drew the conclusion, quite unsubstantiated of course, that the origin of the animalcules could only be ascribed to "eggs, seeds or pre-organized corpuscles which we understand and distinguish by the genetic name of germs."[76] These experiments of Spallanzani, while casting doubt on Needham's "vegetative force," did not repudiate Buffon's concept of "organic molecules." There was no reason why these organic molecules could not be disseminated in the air and act in a way analogous to these eighteenth-century "seeds."

The opposition of Spallanzani and of his colleague Charles Bonnet to spontaneous generation rested on their adherence to traditional eighteenth-century mechanism and the necessary consequence that animal form was inexplicable by ordinary laws of matter and motion. Indeed Spallanzani began his first paper on spontaneous generation with an attack on the concept of *"forces plastiques,"* which, he argued, "has been universally discredited by schools of modern science. . . . They reject these forces as absolutely useless and ineffective. For myself, I am vigorously persuaded that even today one should still deny them . . . and that the system of ovipary should be maintained in all its splendor and brilliancy."[77] Nature was static, not dynamic; spontaneous generation and epigenesis were therefore impossible. To both Spallanzani and Bonnet, the generation of living forms was only possible through fission or the preexistence of germs. As Bonnet remarked in 1764, "A sound philosopher has eyes that discover in every organized body the ineffaceable imprint of a work done at a single stroke, and which is the expression of that Adorable Will that said, 'let organic bodies be,' and they were. They were from the beginning and their first appearance is what we very improperly call generation, birth."[78]

Bonnet, however, recognized variations in nature. His system of preexistence did not imply the existence of an infinite series of preformed individuals. The concept of infinite divisibility, he remarked, "is a geometric truth and a physical error."[79] The male semen, which nourished the egg, was capable of altering the subsequent development of the egg. What preexisted was a regulatory principle, by which the arrangement of the nourishing atoms is determined.

His denial of physical *emboîtement* led him to dismiss the concept that all parasitic worms were created in miniature in the first host organism, a belief that was reinforced by the theological problem posed by the existence of worms in an "innocent" Adam. He was led reluctantly to conclude that the original worm seeds were produced after the "Fall," a belief that was, of course, contrary to Holy Writ. He was also forced to conclude that worm eggs enter the body from the outside, implying the role of chance, which was so antipathetic to eighteenth-century concepts of natural law.[80] Again we see illustrated the great problem posed by parasitic worms to those opposed to spontaneous generation. It was at this time, 1780, that the Copenhagen Prize was offered regarding a solution to worm origins, a prize that was awarded to proponents of spontaneous generation (*see* chapter three). It is significant that their arguments were similar in one respect to those of the adherents of preexistence theories; both groups shared a common abhorrence of chance. It was argued that worms destined by nature to live only inside other organisms could not arrive there by chance; therefore, they must be inborn.

The fact that support for spontaneous generation (of worms at least) could be found among those adhering to the orthodox eighteenth-century view of stability and natural law should not be allowed to obscure the fact that the gradually increasing acceptance of spontaneous generation at the end of the eighteenth century was associated with two outlooks utterly opposed to what lay behind pre-existence theories. The first was the materialism of the French *philosophes,* and the second, utterly in opposition to the first, was the rise of the ideas that were later to coalesce in the writings of the German *Naturphilosophen.* In both instances acceptance of spontaneous generation was essential.

Although much of Cartesian thought became harmonized with a Christian and theological outlook, it is clear also that Cartesian physics led directly to the materialism of late-eighteenth century thought. As Henry More wrote in his *Enchiridion metaphysicum* of 1671: "No greater blow and wound can be inflicted on the most essential part of Religion than by presuming the possible explanation of all phenomena clearly through their causes. As if this material world . . . could have engendered itself! Yet such is the Cartesian hypothesis."[81]

By the middle of the eighteenth century mechanical-deism had fallen into ill repute. Thereafter, despite the opinions of Spallanzani and Bonnet, materialism gained favor, flourishing in Denis Diderot and *L'Encyclopédie.* This school of materialism not only attempted to exclude from the study of nature any reference to final causes and theological ideas, but also put forward a theory of cosmic development based only on the laws of matter in motion. The generative powers of Abraham Trembley's *Hydra,* which seemed to suggest that nature had an innate capacity to form organic beings, and Needham's experimental evidence in support of spontaneous generation became important justifications of the new materialism. As Vartanian observed, spontaneous generation seemed to validate the broadest claims of the *philosophes'* thought—that the causes of all development must be sought for inside, not outside, of nature.[82]

The early Diderot was an orthodox deist who believed that the effects of matter in motion were "limited to developments of what already exists."[83] But, with the suppression of *L'Encyclopédie* in 1759, he became part of a politically radical group of materialists dedicated to the overthrow of the monarchy and the propagation of atheism. Their literary efforts culminated in 1770 with the publication of Paul d'Holbach's *Système de la nature,* and with the appearance of Diderot's *Le rêve de d'Alembert* in the previous year. In this work Diderot had come to accept the existence of spontaneous generation.

D'ALEMBERT: You do not believe, then, in pre-existent germs?
DIDEROT: No.
D'ALEMBERT: Ah! That pleases me!
DIDEROT: Such an idea is contrary to experience and reason: contrary to
   experience because no amount of experience would ever discover these
   germs in an egg or in most animals before a certain age. Against reason,
   because, although there may be no limit to the divisibility of matter in the
   mind, there is such a limit in nature. It is repugnant to envisage a totally
   formed elephant in an atom and in that atom another elephant, and thus to
   infinity.[84]

The germ he explained "is only an inert and thick fluid," which
becomes an organism only through the agency of heat.[85] He expressed support for the experimental findings of Needham, where
"in the drop of Needham's water, everything is executed in a wink of
the eye."[86]

What was the elephant in the first place? Perhaps the enormous animal we see
today, perhaps only an atom, for both are equally possible. Both require no
other suppositions than motion and the various properties of matter.... The
Elephant, this enormous mass, organised, the product of fermentation! Why
not? It is less difficult to relate that huge quadruped back to its original matrix
than to relate the tiny worm to the molecule of flour that gave rise to it. The
miracle is life, the sensibility, and the miracle is a miracle no longer.... Once I
have seen inanimate matter actually become sentient, then nothing should
ever be able to astonish me again.[87]

The impact of such extreme materialistic views on the scientific
content of the spontaneous generation debate was not great. If Descartes' ideas seemed absurd, those of the materialists seemed even
more so. Their impact was indirect, resulting in the association of the
doctrine of spontaneous generation with materialism, atheism, and
political radicalism to a much greater extent than before. *As a result
defense of preexistence and attacks upon spontaneous generation became a
central tenet of the Christian faith.* This was clearly manifested in France,
where the doctrine also became associated with the *idéologues,* whose
teachings were blamed for the horrors of the French Revolution. This
association of spontaneous generation with the revolutionary forces
of atheism and materialism was to have a profound impact on its
subsequent history in nineteenth-century France.

In Germany the doctrine of spontaneous generation became associated with an approach diametrically opposed to that of materialism, namely *Naturphilosophie.* As a result, its subsequent history
differed markedly from that in France.

The roots of this outlook can be traced, in part, to the old alchemical tradition that still prospered in eighteenth-century chemistry. To chemists of that period the complex nature of living organisms implied a basic instability that necessitated the continual presence of a "vital spirit" in order to preserve integrity of organization. Without it, the organism would disintegrate or corrupt. To George-Ernest Stahl, one of the most influential of this chemical school, it was the soul that continually directed living and generative processes. In the realm of animal generation, this spiritualistic tradition found its most influential spokesman in Casper Wolff, whose *Theoria Generationis* appeared in 1759, a work that was vastly admired by the *Naturphilosophen* at the beginning of the nineteenth century. Nature, to Wolff, was not a machine, thus the concept of preexistence was based on an utterly untenable foundation. Growth, change, and development were the very essence of the natural world, and therefore required no causal explanation. Thus, for Wolff, it was necessary only to *describe* the events of embryological development in terms that were, of course, epigenetic. That is to say the process of embryology is not that of *l'évolution* or unfolding of preexisting parts, but a true and gradual process of growth and differentiation. As Gasking points out so clearly, the difference between this explanation and that of the mechanists was not one to be resolved by the "facts" of embryology. The facts were not in question, since the appearance of the parts was compatible with both epigenesis and preexistence. What was at issue here was a fundamental difference in outlook on nature. Nature, to Wolff, was unified by a process of development.[88]

Einheit and *Entwicklung*, unity and development, were the key tenets of *Naturphilosophie*, which became established in Germany during the first three decades of the nineteenth century. The acceptance of epigenesis and spontaneous generation was essential to these beliefs. Hence, it was during this period that the doctrine of spontaneous generation reached its zenith of popularity. To this triumph of spontaneous generation we now turn.

Nullum vivum ex ovo. Omne vivum e vivo.

THREE

THE TRIUMPHANT AGE
OF SPONTANEOUS GENERATION

The first three decades of the nineteenth century witnessed the triumph of spontaneous generation. In these years most of the foremost natural historians came to believe in the spontaneous generation of infusorians and parasitic worms. Much of this popularity rested on the theory's general agreement with empirical evidence, but to a very large extent it found its ultimate expression in the teachings of the German "romantic" philosophers of the early nineteenth century, particularly those espousing the doctrines of *Naturphilosophie*. It was in Germany, far more than in England or France, that spontaneous generation flourished. It was, moreover, this association with the "romantics" which enabled the concept to retain considerable support in Germany up until the 1870s. Not only was Ernst Haeckel's monistic *Weltanschauung*, in which spontaneous generation had a central place, a restatement of the old *Naturphilosophie*, but even the most vociferous opponents of the earlier philosophies sought for a unity in nature which very often led them to a belief in spontaneous generation.

Long regarded with suspicion or outright hostility by historians of science, the *Naturphilosophen* are now seen as having had a very formative influence on early-nineteenth-century German science. Entrenched in the German universities, the *Naturphilosophen* exerted considerable influence in the training of many early figures in German scientific endeavors.[1]

Naturphilosophie rested on the work of Friedrich Schelling, who in turn followed the earlier writings of Immanuel Kant and Friedrich Jacobi. Their main concern was with the problem of knowledge, and more specifically with the relationship between *Verstand* and *Vernunft;* the first, the faculty of "understanding," constituting the "lower" method of the sciences, and the second, the faculty of "reason," being the method of philosophical inquiry.[2]

The distinction between these two modes of thought had formed the basis of Kant's *Kritik der reinen Vernunft* (Critique of Pure Reason), published in 1781. At an earlier period in Kant's life, he had believed that everything in nature was completely deterministic and all philosophical concepts derivable from experience. On such a basis he had produced his famous *Allgemeine Naturgeschichte und Theorie des Himmels* in 1755. Later he arrived at the same position as David Hume concerning the problem of causality: that experience alone cannot justify a belief in universal causality and thereby a belief in the uniformity of nature. Kant extricated himself from this dilemma by what he came to call the "Copernican Method" of critical philosophy. He claimed that, since our perception of the universe is due to our peculiar position on a revolving planet, likewise our perception of space and time is due to our method of sense perception. Extending this further, he concluded that, through sense perception alone, we cannot know of things-in-themselves; that our mind, interposed between the real world and ourself, interprets the chaotic plurality of sense data and synthesizes them to produce the unity of natural laws. This is the only knowledge available to us when following the empirical methods of the natural sciences, and such knowledge cannot tell us anything of the absolute, immutable reality of the world in which we live. Concepts of time, space, magnitude, and causality are therefore subjective, and a sharp distinction exists between the real world and the world of experience, and between man and nature.

Friedrich Jacobi could not accept this limited method of knowing. For him, man had another faculty, reason, or *Intellektuelle Anschauung,* which alone gave man the ability to know the real essence of the universe. Similar views had been put forward by Kant so that Thomas Carlyle could claim, in 1827, that: "Reason, the Kantists say, is of a higher nature than the Understanding; it works by more subtle methods, on higher objects, and requires a far finer culture for its development . . . for Reason discerns Truth . . . while Understanding discerns only relations."[3]

The superiority of reason over understanding was the theme of Schelling's confused and obscure teachings in *Naturphilosophie.* Through an inward-looking self-knowledge, the real essence of being is achieved directly. Knowledge is derived, he wrote in 1795, "from an experience immediate in the strictest sense of the word, an experience that is self produced and independent of all external causality."[4] What one came to know by this method remains forever a mystery, except that it seemed to have had the absoluteness of a Godlike essence, but an essence that one arrived at by self-intuition rather than by faith.

These ideas, seemingly so antithetical to the scientific enterprise, exerted a considerable influence when Schelling incorporated the concept of a "pure unity of being" into the plurality of things perceived through sense experience. In other words, he claimed, in his

Von der Weltseele of 1798, the concept of the "One" was necessary even for the natural sciences. No longer asserting that the "One" was the sole reality, he claimed instead that it was "the primary cause of all these changes and the ultimate ground of all nature's activity."[5]

This incorporation of transcendentalism into the realm of science led to two important concepts that came to dominate this period, *Einheit* and *Entwicklung.* The latter concept of "development" (*Entwicklung*) defines the world in terms of a creative process gradually being manifested over time, or, as Arthur Lovejoy termed it, "Progressionism without Transformism."[6] The concept of "unity" (*Einheit*) was expressed by the *Naturphilosophe* Carl Carus in the following terms: "We hold, therefore, that the true end of scientific enquiry is not to define and demonstrate the highest principle, but to trace other truths up to this, to show the harmony which exists between nature and mind, or to discover a unity of law in the multiplicity of phenomena."[7]

The concept of unity had a profound influence on the spontaneous generation debate. Implicit in the concept lay the denial of any distinction between living and nonliving entities and even of death itself. Equally significant, the total universe was seen to be essentially organic. Life was defined in terms of the unity and totality of all processes occurring in a living being, "a constant manifestation of an ideal unity through a real multiplicity." Paralleling the way in which the totality of individual intellectual intuitions resulted in the knowledge of the single absolute "One" and not of a multitude of specific individual "Ones," the totality of individual lives of individual organisms led to the conclusion that there was a unity of life in all of nature, rather than a plurality of individual lives. "Thus we find in fact the idea of life, that is, the constant manifestation of unity through multiplicity, exhibited by universal nature; and are therefore bound to consider nature collectively as one vast and infinite life . . . an absolute and proper death is inconceivable."[8]

Since a living spirit or *Geist* pervaded all of nature, it followed that all parts of nature were part of a greater living whole. Acceptance of spontaneous generation was thus a necessary ingredient of *Naturphilosophie,* whether it were the generation of an organism from unorganized matter or from matter in a "living" or "dead" organism. Both these processes were identical, any distinction between heterogenesis and abiogenesis being meaningless. In the same way, the *Naturphilosophen* accepted the ideas of epigenetic embryological development since this implied that the seed contained ideally within itself the whole animal. In essense the belief in spontaneous generation and the belief in epigenesis were one and the same concept. The German embryologists at this time rejected the older ideas of preexistence and its associated denial of spontaneous generation and accepted both epigenesis and spontaneous generation.

Naturphilosophen, such as Carus, Gottfried Treviranus, Lorenz Oken, Karl Ernst von Baer, and Carl Burdach, all accepted the continuous occurrence of spontaneous generation. The *Biologie* of Treviranus included, for example, a description of the spontaneous generation of infusorians from exposed spring water.[9] Oken's *Lehrbuch der Naturphilosophie* is filled with aphorisms dealing with spontaneous generation:

All generation is *Generatio aequivoca,* whether imparted by sexes or not. For the generative juices of the sexual organs themselves are nought else than organic primary mass, and have originated by division.

Death is no annihilation, but only a change. . . . Death is only a transition to another life, not unto death.[10]

Two aspects of spontaneous generation were developed by the German biologists at this time. The first involved studies in epigenetic development visualized as a spontaneous generation; the second was a series of important studies dealing with parasitic worms initiated by the great parasitologist, Carl Rudolphi. Although not a *Naturphilosophe,* Rudolphi and his followers made reference to concepts developed by these philosophers. Curiously, both embryologists and parasitologists based their work on earlier writers of the eighteenth century. The former looked back to Casper Wolff, the latter to Marcus Bloch and Johann Goeze, winners of the Copenhagen Prize.

The arguments which had been used by Bloch in defense of spontaneous generation were very typical of eighteenth-century thought. They rested on the abhorrence of chance: that, if it could be proved that parasitic worms live only as parasites and not also as free-living forms, then they could not reach their hosts by the fortuitous meeting of host with parasite. He presented twelve "proofs" to illustrate that worms "are destined by nature to live only in other animals."[11] Clearly these proofs rested on a great many accurate observations, far surpassing anything produced previously in the eighteenth century. The proofs are:

1. Worms are never located outside any animal body.
2. Worms are sometimes located in young animals, recently born animals, and even in abortions.
3. Many worms, which are never found in the gut, occur in the interior parts of the animal that are without any passage to the exterior.
4. Worms live in locations where other organisms are digested.

5. Worms not only live in an animal body, but thrive and multiply there.
6. Worms rapidly die after removal from the host body.
7. Most animals have their own particular parasitic worms.
8. Worms have a specialized structure which enables them to live only in an animal body. They not only carry special organs, such as hooks and suckers, but are devoid of other structures, having no need for them.
9. Worms produce a great quantity of eggs, since they are unable to deposit these eggs like other animals and need to produce excess numbers to replace those lost through the excrement.
10. There are a greater number of female than male worms present, a necessity because so many eggs are wasted.
11. Worms cannot live equally well in all animals.
12. Worms do not always cause illness.

Having shown that the worms were not accidental inhabitants of the host animal, but were perfectly fitted for life in that environment, he went on to prove "by reasoning and facts" that the seeds of such animals must be inborn. The arguments he used were based on the assumption, that animals always laid their eggs in a suitable place for development. This fact was well documented through work on insects that had dominated eighteenth-century natural history. The Creator, he argued, could not allow the eggs of animals destined to live in a certain locality to arrive there by chance. He substantiated these claims further by explaining that, if eggs were received from external causes, then all animals in the same locality and utilizing the same food would necessarily be infected with the same worms. Since this was manifestly not the case, the eggs must be generated inside the host animal. It is worth mentioning that this problem of parasite distribution was identical to the problem of disease susceptibility: that, in both cases, the contagious nature of infection was denied. In the former case, spontaneous generation was inferred, in the latter, the concept of "miasmata" usually prevailed.

Bloch concluded, in typical eighteenth-century fashion, that a moral lesson must be learned from the existence of the parasitic worms. They were necessary to establish the harmony of the whole, to show that every living creature must inevitably serve as food for another. "These same worms, I say, seemed to be placed expressly in our interior, to prove that we are destined to nourish animals in our turn, as they nourish us."[12]

Although Bloch's twelve proofs were simply set out to show that worms were obligatory parasites, many of them were taken up and elaborated on by the German parasitologists of the early nineteenth century to show that the adult parasites were products of spontaneous generation. The pioneer among these early parasitologists was Carl Rudolphi, teacher of the great Johannes Müller. His main concern was with descriptive and classificatory studies of the various worms, and his famous text of 1819, *Entozoorum Synopsis,* listed nearly a thousand species of intestinal worms. His importance in the controversy rests on the fact that he erected one group, the Cystica, which included, among others, the forms that we now know to be the larval stages of the adult tapeworms, the Cestoidea. It was the Cystica which were to become the most cherished and obvious examples of spontaneous generation during the next forty years.

The problem of the origin of parasitic worms was the subject of *Über lebende Würmer im lebenden Menschen* (1819) by Johannes Bremser, a close friend and colleague of Rudolphi. Bremser showed, by utilizing nearly all of Bloch's earlier arguments, that the worms were obligatory parasites incapable of living outside the human body. He concluded from this that there were only three feasible explanations to account for their origin. These may be summarized as follows:

1. By passage of the eggs or worms from one host to another through air, food, or water.
2. By passage of the eggs or worms directly from one or the other of the host's parents. Either:
    a. By passage through the sperm,
    b. By passage through the egg,
    c. By direct passage through the placental barrier into the fetus, or
    d. By passage through the mammary glands during lactation.
3. By direct spontaneous generation of the worm from the tissues of the host.

Bremser's long and detailed examination of these various hypotheses leaves no doubt that the evidence then available thoroughly supported the doctrine of spontaneous generation. He argued against the first hypothesis by reference to the curious distribution of the two common human tapeworms: that, whereas the inhabitants of Russia and Poland are infected with *Dibothriocephalus,* the remainder of Europe is annoyed by *Taenia.* Neither could it be explained, on the basis of the first hypothesis, how worm eggs were passed among animals, like man, which eat neither each other nor each other's feces! He drew attention to the seeming anomaly that herbivores were as susceptible to tapeworms as carnivores and that the former were more prone to infestation by Rudolphi's Cystica (such as hydatid cysts) than the latter. He remarked also on one of the rare feeding experiments made before 1851 that seemed to deny the validity of the first hypothesis. In 1806, Schreiber had fed skunks with worms, worm eggs, milk, and bread crumbs for six months without finding any trace of parasitic worms. All these phenomena, of course, made sense only after the discovery of intermediate hosts. Before these discoveries were made, passage of eggs through air, food, and water had to be denied.

Bremser dismissed the second hypothesis with contempt, since it implied either that parasites were present at the beginning in the body of the first created individual of the host species, or that the modern worms had become transformed from the original creation of a few worms. The former was clearly absurd since the number of known species was then so great that poor Adam must have been literally crawling with them. The latter he dismissed by suggesting that anyone capable of believing that hydatids, nematodes, and tapeworms all had the same origin, "would not, in consequence, regard a man to be devoid of his senses who believed that any animal, an elephant, for example, would after a time become the father of whales, lions and kangaroos."[13]

There seemed to be only one rational answer, and most parasitologists agreed with it: "They must owe their existence to a spontaneous generation. We cannot find cause for the first production of helminths other than in a change in the nature and constituents of those substances necessary for the general nourishment of the body." Disease and parasitic worms were produced by a disharmony in the body. When the gut contained more animalized material than it could absorb, then such material transformed itself into a parasitic intestinal worm.

Disease is produced by the disharmony of the function of different organs. A similar disharmony must exist if worms are to be generated. If in the stomach, from the ingested food no other nourishment is prepared than what is necessary for the replacement of the excreted material or for the enlargement and growth of the body, no worms will be generated. If . . . more substance is animalized than can be absorbed, we find nothing easier than this.[14]

Bremser supported this belief in spontaneous generation by metaphysical speculation, which showed the stamp of early-nineteenth-century German philosophical concepts, and which, in many cases, was very similar to the views of the *Naturphilosophen*.

The formless mass of the primitive earth was "animated, that is to say endowed, with a living Geist," which "did not yet consist of isolated bodies."[15] Individual organisms appeared with the precipitation of the primary stratum, after which a series of catastrophes, annihilations, and recreations culminated in the creation of man, himself the final cause of these activities. Life was not to be found in a mixture of various substances: "that would be crude materialism. The primary cause of life lay in the Geist."[16] And thus, in words that almost mirror those of Oken, Bremser wrote: "Death does not occur in the whole of the organized nature. Death is only a transition to a new life, to another form of life."[17] This belief in the transition of life forms led him to the statement in support of spontaneous generation:

I take as proven the spontaneous generation of moulds and infusorians from dead organic bodies, but if independent living bodies are generated from dead organic bodies, such generation must take place more easily in living organisms themselves . . . we can assume that organisms generated from living bodies must be more perfect than those generated from dead ones . . . as the primary cause of life, the Geist, is increased and acts more powerfully. From dead plant and animal organisms sometimes moulds, sometimes infusorians are generated. . . . In living organisms the new product always corresponds to the nature of the organism from which it is generated. Lichen and moss grow from plants; intestinal worms, lice and mites on animals.[18]

He justified his views on spontaneous generation further by reference to the unity of the cosmos, and to the belief that man and his worms paralleled the universe around him: "Whenever the earth changed its form of existence, the existing creations were also destroyed. The same thing occurs to the worms; when the host animal dies they are also destroyed."[19] In a similar vein, he saw a unity in the variety of reproductive types found in parasitic worms, ranging from simple budding in hydatids and coenuri, through hermaphroditic and androgynous flukes and tapeworms, to the dioecious nematodes, all of which were produced initially by spontaneous generation:

The different modes of generation that one sees in parasitic worms prove that the progress of creative nature on a minute scale is identical to the progress on a grand scale, and that it does not operate any differently in living animals than in our living globe. In other words, in intestinal worms one finds a repetition of all the types of generation which exist in the animal series.[20]

Bremser was not a *Naturphilosophe* as he himself insisted. He maintained, for example, that there was a fundamental difference between brute and animate matter and that a living organism could never be produced out of brute matter. He furnishes a classic example of a belief in heterogenesis and a denial of abiogenesis, a distinction which is completely meaningless in *Naturphilosophie*. Nevertheless, there were parallels between certain of his speculations and those of the *Naturphilosophen*. His association of heterogenesis with such concepts as *Geist* and the micro-macrocosm did lead a later generation of German physiologists to associate heterogenesis with speculative vitalistic biology which they found so abhorrent.

In France, support for spontaneous generation never reached the heights that it achieved in Germany, but in the early decades of the nineteenth century it did reach a level of popularity which it never enjoyed again. Naturally, Frenchmen like Étienne Geoffroy Saint-Hilaire, who came under the influence of *Naturphilosophie*, accepted spontaneous generation, as did his pupil, Antoine Dugès.[21]

The major figures in the French drama were Pierre-Jean-Georges Cabanis and Jean-Baptiste Lamarck on one side, and Georges Cuvier on the other. Cuvier was a giant among French scientists whose influence extended well into the middle of the nineteenth century.[22] He stressed the harmony and functional teleology of every organism: that every part had a function which served the well-being of the whole. It was inconceivable to him that such an integrated whole could ever be produced by a simple aggregation of matter. "Life has always arisen from life. We see it being transmitted and never being produced."[23] The functional integrity of a living organism and the necessity of retaining this integrity led Cuvier to accept preformationist rather than epigenetic doctrines of development. The latter hypothesis, which saw development in terms of a simultaneous growth and differentiation, could produce only physiological chaos in the eyes of Cuvier. He was therefore committed to a belief in the preexistence of germs and the corresponding denial of any form of spontaneous generation. Cuvier's greatest imput, as far as later arguments over spontaneous generation were concerned, was his overthrow of the "unity of type" concept and its replacement by four *embranchements*. This was significant because, through Lamarck, spontaneous generation became associated with unity of type and evolutionary beliefs.

From his law of the subordination of characters Cuvier came to see the nervous system as the most important in the animal organization. He concluded, on the basis of the structure of the nervous systems, that "there exist four principal forms, four general plans, upon which all of the animals seem to have been modeled."[24] These were the Vertebrata, Mollusca, Articulata, and Radiata. This scheme was incorporated into his *Règne animal*, first published in 1817 and republished in 1830.[25] Cuvier was thus opposed to the Great Chain of Being concept, and according to William Coleman he was also opposed to any idea of an hierachical chain within each of the separate *embranchements*. He was also opposed to any temporalization of the Chain of Being, whether by German *Entwicklung* or Lamarckian transformations. These views led Cuvier into his famous controversies with Geoffroy Saint-Hilaire and Lamarck.

Between 1799 and 1800 Lamarck came to believe in species mutability.[26] His belief was intimately associated with his interest in the problem of extinction. The fact that fossil remains had no living counterparts did not lead him to accept that they were remains of extinct forms. On the contrary he assumed that: "These species have changed in the course of time and that presently they have different forms from those of the individuals whose fossil remains we find."[27] At the same time he came to believe in spontaneous generation.

One of the crucial factors which led Lamarck to accept spontaneous generation was a change in outlook, which took place in the 1790s, over the nature of life. In 1778 he accepted that organic beings possessed a "very well marked vital principle and the faculty of reproducing their own kind."[28] By 1797 he had modified these views and denied the existence of a vital principle, although he still accepted that "everything that has life comes from an egg."[29] But by 1800 he had come to accept spontaneous generation.

He regarded all natural phenomena, inorganic and organic, to be a result of the interaction of two forces: the universal force of attraction and the repulsive force between imponderable matter. Such views were common during this period when imponderable matter was used to explain many physical phenomena such as heat or electricity. These forces alone produced vital activity: "Nature has no need for special laws, those which generally control all bodies are perfectly sufficient for the purpose." Such forces produced degradation in inorganic bodies, and vital activity in living bodies. On numerous occasions Lamarck stressed that "the cause, which produces vital forces in bodies whose organization and structure permit those forces to exist and excite the organic functions, could not give rise to any such power in crude or inorganic bodies."[30] Vital activity was therefore a product of organization which, being absent in inorganic bodies, could not manifest such activity. It was this difference of organization, rather than any vital principle, which allowed Lamarck to speak of "a great hiatus" between living and inorganic bodies.[31]

Despite this "hiatus," however, Lamarck clearly was impressed by the simplicity of primitive organisms at the base of his chain of degradation. His belief that simple infusorians "are only very tiny, gelatinous, transparent, contractile and homogeneous bodies"[32] distinguishes him from most eighteenth-century naturalists who stressed the complexity of all living things and thus refuted the doctrine of spontaneous generation. Lamarck, on the other hand, struck by the emphemeral nature of those creatures, which constantly reappeared each year after destruction by harsh environmental conditions, was able to conclude that they arose by a continuous spontaneous generation. He admitted that it was difficult to understand how this came about, but suggested it was due to the penetration of imponderable matter into nonliving materials. The nature of this nonliving material is not clear from Lamarck's writings; indeed, he never really posed that question. Since he believed that mineral material was a product of organic degradation, the source of life appeared to have been material that was once alive: "The body that is most fitted for the reception of the first outlines of life and organization is any mass of matter apparently homogenous, of gelatinous or mucilaginous consistency, and whose parts though cohering together are in a state closely resembling that of fluids, and have only enough firmness to constitute the containing parts."[33]

The origin of the simplest organisms from this mucilaginous material involved simply "filling these little cellular masses with fluids and in vivifying them by setting these contained fluids in motion by means of stimulating subtle fluids which are incessantly flowing in from the environment."[34]

When these concepts of fossil mutability and spontaneous generation were added to his belief that each kingdom of nature could be arranged into a single hierachy of increasing complexity, then Lamarck came to postulate his theory of evolutionary change. This process may be likened to a moving escalator. Simple forms of life were fed into the bottom of the staircase by a continuous spontaneous generation and thereafter transformed themselves into more and more complex and perfect forms through the agency of a progressive driving force. By 1815, however, his conviction that parasitic worms were spontaneously generated led him to modify his views to some extent.

Lamarck was initially reluctant to admit that worms were spontaneously generated in the body of the host. In his *Système des animaux sans vertèbres,* published in 1801, while admitting that worms "grow, live and multiply everywhere in the bodies of other animals," he concluded, nevertheless, that "one must not say from that, that these worms are born in the same animals in which they exist."[35]

At this time Lamarck believed that spontaneous generation was unique to the infusorians and thus could only occur at the bottom of the chain. Spontaneous generation of parasitic worms would then be "contrary to the known march of nature." But, by 1809, he spoke of the animal chain as commencing in two places, the primitive infusorians and the worms, a concept that was further elaborated in his text

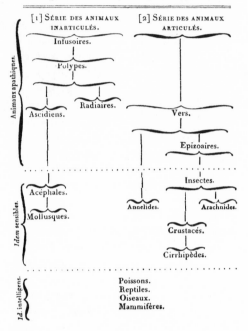

*ORDRE présumé de la formation des Animaux,
offrant 2 séries séparées, subrameuses.*

Lamarck's double animal chain, each commencing with a spontaneously generated form. From his *Histoire naturelle des animaux sans vertèbres* (Paris, 1815).

of 1815, *Histoire naturelle des animaux sans vertèbres*. Curiously, the parasite branch seems to have become the more important, for the worms gave rise to the articulates and the vertebrates. The infusorial branch led to polyps, radiarians, ascidians, and finally the molluscs. In addition, it was only the parasite branch which was "capable of giving rise to feeling."

The French, through the writings of Lamarck, came to associate the doctrine of spontaneous generation with theories of transformism. Indeed, so strong was this association that they later refuted Darwin's *Origin of Species,* mainly on the grounds that it implied a continuous spontaneous generation.[36] Equally significant to the spontaneous generation debate in France, they also associated the doctrine with the writings of the late-eighteenth-century *philosophes* and the *idéologues* of the revolutionary period, among whom was Pierre-Jean-Georges Cabanis. Cabanis, a materialist who believed that "all phenomena of the universe have been, are, and will always be the result of properties of matter," believed like Lamarck that infusorians and parasitic worms arose by a spontaneous generation.[37]

This association with transformism, materialism, and the

idéologues resulted in the politicization of the spontaneous generation controversy in nineteenth-century France. Not only were the *philosophes* blamed for generating ideas leading to the chaos of the French Revolution, but *idéologique* principles in thought and practice were fundamentally opposed to Catholicism. Heavily involved with the educational reforms of the revolution, the *idéologues* were instrumental in replacing the theologically orientated goals of education of the *ancien régime* with the secular and republican ideals expressed in the Law of the Third Brumaire (1795). Education became an instrument of republican civicism, founded on reason and analysis and directed against prejudice, superstition, and the study of religion.[38] Not surprisingly French Catholics and other opponents of republicanism identified *idéologie* with atheism. Napoleon's successful association with the Catholic Church in opposition to atheistic revolutionaries and his public refutation of the *idéologues,* on whom he placed much blame for the *terreur* of the Revolution, linked the doctrine of spontaneous generation with forces opposed to the stability and order of the French State. Moreover, this political and theological opposition to spontaneous generation received scientific justification from Cuvier's ridicule and opposition to Lamarckian transformism. Faced with such overwhelming opposition, the doctrine enjoyed only a short period of popularity in France, so that long before the time of Pasteur the doctrine was generally unpopular.

Despite these overtones, however, a few noted French naturalists such as Bory de Saint Vincent, Charles-Louis Dumas and Jacques-Armand Deslongchamps continued to accept the validity of spontaneous generation. One sees acceptance of spontaneous generation in such publications as *Dictionnaire classique d'histoire naturelle* of 1825 and *Dictionnaire des sciences naturelles* in 1828. It appears it was accepted without much enthusiasm simply on the basis of empirical evidence. There is certainly no indication that this acceptance stemmed from a prior commitment to any metaphysical system.

In Britain, eighteenth-century abhorrence of spontaneous generation was carried forward well into the nineteenth century. Completely indifferent to the strong arguments mounted in its favor, British naturalists continually claimed that it was "unphilosophical,"[39] "contrary to an unerring law of nature,"[40] and worse, "atheistic."

This theological opposition to spontaneous generation was dramatically illustrated by early work on plant photosynthesis. In 1772 Joseph Priestley had concluded that plants "reverse the effects of breathing" by removing phlogiston from air made noxious by animals breathing or putrefying in it.[41] Being unaware of the significance of light to this process, he later had difficulty repeating his earlier findings. In June, 1778, he removed his plants from the phials and was astonished to note that the water continued to yield pure air and that the sides of the phials were covered with a "green kind of matter." At this stage he concluded that the green matter "can neither be of animal or vegetable nature, but a thing *sui generis*."[42] Although this first conclusion was based on its presence in corked phials from which "external air, or animalcules" had been excluded, he quickly changed his opinion on the basis that the green matter did not appear in fresh distilled water with no communication with outside air.[43] In his earlier paper, however, he concluded that "when water is brought into a state proper for depositing that green matter, it is, by the same process, prepared for the spontaneous emission of a considerable quantity of pure air."[44]

John Ingenhousz viewed Priestley's findings with religious ecstasy. "We are assured," he wrote, "that no vegetable grows in vain," that, whereas in the winter, water alone can purify air, in summer, "that immense quantity of animal substances, and many others, which undergo putrefaction by the warmth of the weather, seems to require an additional power or agent to counteract it: and this office is destined in the leaves."[45]

Another such agent was the green matter, "seemingly vegetable," produced spontaneously from the water, indicating the tenacity of a belief in the transmutability of water, springing from earlier iatrochemical writings. "I should rather incline to belief, that this wonderful power of nature, of changing one substance into another and of promoting perpetually that transmutation of substances, is carried on in this green vegetable matter in a more ample and conspicuous way. The water itself, or some substance in the water, is, as I think, changed into this vegetation."[46]

Although Ingenhousz claimed that such occurrences illustrated "that this great universe is not the offspring of chance, . . . but that it has been made by an Omnipotent Being,"[47] Priestley held other views on the issue. Believing in the preexistence of germs, he considered spontaneous generation not only to be impossible, but worse, to be a phenomenon which lacked an intelligent cause. Such a belief is, he wrote, "unquestionably atheism. For if one part of the system of nature does not require an intelligent cause, neither does any other part, or the whole."[48]

Priestley here was responding to the infamous poem of Erasmus Darwin, *The Temple of Nature,* in which the green matter had been mentioned as having "but a spontaneous origin from the congress of the decomposing organic particles."

> Then, whilst the sea at their coeval birth,
> Surge over surge, involved the shoreless earth;
> Nursed by warm sun-beams in primeval caves
> Organic life began beneath the waves. . . .
> Hence without parent by spontaneous birth,
> Rise the first specks of animated earth.[49]

From a series of rather crude experiments with open flasks, flasks of pump water covered with olive oil, flasks with ground-glass stoppers, flasks inverted in a vessel of mercury, and flasks with loose tin covers, Priestley concluded that the results "are perfectly agreeable to the supposition that the seeds of this small vegetable float in the air, and insinuate themselves into water of a kind proper for their growth, through the smallest apertures."[50] He also denied the spontaneous generation of parasitic worms on the invalid grounds that "most of these very worms have been found *out* of the body."[51]

Clearly, however, Priestley's opposition to spontaneous generation rested mainly on theological arguments so typical of British natural theology:

But these microscopic vegetables and animals, there is every reason to think, have as complete and exquisite an organic structure as the larger plants and animals, and have as evident marks of design in their organization, and therefore could not have been formed by any other decomposition or composition of such dead matter, whether called organic or not, without the interposition of an intelligent author.[52]

Despite British hostility, spontaneous generation was widely accepted during the first three decades of the nineteenth century. It was associated with *Naturphilosophie* and the German philosophical traditions, with late-eighteenth-century materialistic ideas, and with the new concept of progressive transformism put forward by Lamarck. In addition—and it is impossible to overemphasize this fact—it was strongly supported by empirical evidence. Both Burdach, the German *Naturphilosophe,* and Cabanis, the French materialist, pointed this out very clearly:

Those who defend the doctrine of spontaneous generation do so through experience; when one sees an organized being born without being able to discover either a germ, or any way by which the body had been able to reach the place of formation, one admits that nature has the power to create an organized being with heterogeneous elements. One's antagonists seek to establish the probability that there are concealed germs, because they believe these germs necessary. Nature, in their opinion, has only the ability to conserve organized beings, not to create them.[53]

And Cabanis expressed it with these questions:

Why do we think it necessary to admit the existence of supposed corpuscles which one can neither understand nor make perceptible? Why do we believe this vague word, germ, explains the most important phenomenon of nature?[54]

The idea that Redi and Spallanzani had reduced the concept of spontaneous generation to a waning archaic doctrine is manifestly absurd. The first three decades of the nineteenth century were the years in which spontaneous generation became firmly established in the biological tradition. It is not surprising that its demise should be a long and involved process extending over the following century.

FOUR

DOUBT AND UNCERTAINTY

THE INTERMEDIATE YEARS

1830–1859

The period from about 1830 to 1859 was probably the most confused of any since the debate over spontaneous generation began in the middle of the seventeenth century. In an earlier paper dealing with the parasitic worm problem I wrote:

In the events surrounding the rejection of spontaneous generation one can distinguish between those which led to the realization that spontaneous generation was not a satisfactory model and those which led to the construction of a new model. The events of the mid-nineteenth century were unique in that for the first time a new explanatory model became available at the very moment when spontaneous generation was once again being questioned.[1]

This was true as far as parasitic worms were concerned and during this period the doctrine did receive a severe blow with the loss of the parasitic worms as examples of organisms produced by spontaneous generation. But it was not true for microscopic organisms; despite rising hostility to the spontaneous generation concept, there simply was not available a more satisfactory explanation to account for their origins. There was no theoretical reason to believe that these simple organisms could not be produced *de novo* until the model of nuclear continuity was established in the late 1870s. Before that time opinions on spontaneous generation very often were determined by metaphysical assumptions about the nature of life, whether or not such assumptions were substantiated by experimental evidence.

It is clear, however, that by the end of this period the fortunes of spontaneous generation had waned considerably. Never popular in England and by this time unpopular in France as well, the doctrine then received considerable criticism in Germany. Much of this opposition stemmed from the widespread revulsion against *Naturphilosophie*, manifested most clearly among the students of Johannes Müller, the great German physiologist and teacher.

In 1827 Müller consciously rejected his earlier beliefs in *Naturphilosophie*. Still, he remained opposed to the idea that the only difference between organic and inorganic matter lay in complexity of organization. A vital principle was involved, but, he remarked: "Whether this principle is to be regarded as an imponderable matter, or as a force or energy, is just as uncertain as the same question is in reference to several important phenomena in physics."[2] Given the vitalistic nature of his beliefs, he could only accept that spontaneous generation occurred in "animate matter till then disorganized," remarking that it "is still best supported by the facts regarding the Entozoa."[3]

Müller's pupils, the Physicalists, rejected vitalistic Naturphilosophie, and "reacting against the past utilized physics and chemistry to repudiate the earlier German biological traditions."[4] In 1847, three of his students, Hermann von Helmholtz, Emil du Bois-Reymond, and Ernst Brücke united with Carl Ludwig to form the Physical Society. These four "imagined that we should constitute physiology on a chemico-physical foundation, and give it equal rank with Physics."[5] They took as their guide this statement of von Helmholtz: "Physiologists must expect to meet with an unconditional conformity to the law of the forces of nature in their inquiries respecting the vital processes; they will have to apply themselves to the investigation of the physical and chemical processes going on within the organism."[6]

Their initial reaction was to oppose heterogenesis because of its association with vitalism and the teachings of the despised Naturphilosophen. Not surprisingly therefore, many of them, notably Theodor Schwann, von Helmholtz, and Theodor von Dusch, who was a pupil of Müller's student Jakob Henle, became involved with the controversy over putrefaction and fermentation which played an increasingly important role in the spontaneous generation debate.

It was well known that putrefaction was associated with the appearance of microscopic organisms. The question was whether such organisms were the cause of putrefaction or merely a result of it. If the latter were the case, then it could imply that such organisms were produced by the process of heterogenesis. In 1836, Schwann, one of the most influential of the early physicalists, read a small communication to the Versammlung der Naturforscher und Ärzte, in which he argued that boiled meat in an enclosed glass sphere did not putrefy nor produce infusorians.[7] With the publication of his more famous paper a year later, Schwann also became involved in the controversy over fermentation, postulating that the process was caused by yeast, a living fungus.[8] In 1843, Helmholtz confirmed Schwann's findings, although he was later to conclude that putrefaction was a process of chemical decomposition which did not necessitate the presence of any living organisms.[9]

At the same time the spontaneous generation controversy entered into the age-old problem of fermentation.[10] Before 1838 fermentative changes always occasioned a chemical explanation, an explanation which was to reach its apogee in the work of Justus von Liebig. Antoine-Laurent Lavoisier can be considered the originator of these nineteenth-century chemical theories. In the *Elements of Chemistry* (1789), he analyzed the substances involved and maintained that the carbonic acid, alcohol, and acetic acid produced were derived directly from the sugar. "The effects of the vinous fermentation upon sugar is thus reduced to the mere separation of its elements into two portions; one part is oxygenated at the expense of the other, so as to form carbonic acid, whilst the other part, being disoxyginated in favour of the former, is converted into the combustible substance alkohol."[11]

Although generally regarded as a mere chemical agent, the nature and role of the organic ferment, yeast, remained obscure; as also did the relationship between vinous and brewer's yeast. Gay Lussac considered the ferment in grape juice to be formed by the action of air on the juice, but considered brewer's yeast to be of a different nature.

Liebig held that yeast was "a compound of nitrogen in the state of putrefaction or decay ... capable of imparting the same condition of motion or activity in which its atoms are, to certain other bodies."[12] In that way alcohol and carbonic acid were formed by the breakdown of sugar, while yeast, the initiator of such activity, was formed from the nitrogen or gluten constituent of the grape juice.

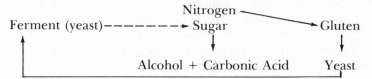

Such a chemical explanation was also utilized to explain putrefaction and miasmatic diseases. The poisons not only caused the putrid changes and disease symptoms but also induced the production of new "morbid viruses" in exactly the same way that yeast induced the production of more yeast. In fermentative changes, no new yeast was produced without the presence of nitrogen. So, in diseases and pathological growths, the diseases would not be contagious and the growths would be benign if only one substance was present. Given the sheer elegance and beauty of this theory, it is not surprising that it enjoyed a widespread and long-lasting popularity.

In 1837 and 1838, three independent workers concluded that

brewer's yeast was not a chemical ferment, but a living organism. On the basis of microscopical examinations, Charles Cagniard-Latour and Friedrich Kützing came to the conclusion that:

The beer yeast . . . is a mass of small globular bodies capable of reproduction, and consequently organized and not a simple organic or chemical substance. That these bodies appear to belong to the plant kingdom and reproduce in two different ways. That they seem to cause the decomposition of sugar only when they are alive, from which one can conclude that it is very probable that the release of carbonic acid and the production of alcohol from this decomposition is a result of their growth.[13]

Schwann, on the other hand, approached the problem from a different point of view. He was concerned, for reasons already discussed, with showing that the organisms associated with putrefaction were not produced by a spontaneous generation. In the two reports already mentioned, he had demonstrated that boiled meat enclosed in a glass sphere did not putrefy even after exposure to previously heated air. Then, wishing to show that this heated air had not been affected in any deleterious way, he allowed it to come into contact with previously boiled cane sugar mixed with brewer's yeast. Believing, presumably, that the ferment was derived by the action of air on the liquid, he was surprised to find that the liquid did not ferment. Rather than concluding that the air had been affected in some way, he assumed instead that germs of molds and infusoria in the air were responsible for fermentation in the same way that he had previously attributed putrefaction to them. Thus he examined the ferment and noted chainlike granules, concluding that yeast was an "articulated fungi" or *Zuckerpilz*. He decided from this, again without legitimate proof, that "alcoholic fermentation must be considered to be that decomposition which occurs when the sugar fungus utilizes sugar and nitrogen-containing substances for its growth, in the process of which the elements of these substances . . . are preferentially converted into alcohol."[14] Naturally Liebig viewed such arguments with justifiable scorn, much of which stemmed from his desire to remove such vague biological descriptions from the new and rigorous science of organic chemistry.

By the time Louis Pasteur arrived on the scene at the end of the 1850s, there was a strong but minority position that yeast was a living organism. Such a view immediately opened up the question of its origin, for the chemical explanation could still be maintained if the yeast was a result, not a cause, of the fermentative processes. Thus, to hold to the chemical explanation and admit that yeast was a living organism immediately led to accepting its spontaneous generation in the fermenting liquid. The result was that those who believed yeast was a living organism responsible for the initiation of fermentation joined forces with those opposed to spontaneous generation. Their goal was to show that yeast was derived from the air and that fermentation could not occur if yeast was removed from the air before contact with the fermentable liquid. Proving this would not only show that yeast was the causative agent of the fermentation, but would also throw fresh doubt on the doctrine of spontaneous generation.

Thus, the problem of spontaneous generation became associated with the problems of fermentation and putrefaction, and there followed the famous experiments of Heinrich Schroeder, von Dusch, and finally Pasteur and Felix Pouchet.

In 1854 Schroeder and von Dusch modified the earlier experiments of Schwann by allowing the incoming air to pass through a filtration tube loosely filled with previously-heated cotton wool.[15] By this means, boiled meat in water was preserved for twenty-three days. On the other hand milk or dried meat did putrefy. They concluded that, whereas the former substances required the addition of an airborne factor in order to putrefy, the latter decomposed spontaneously. In later work Schroeder heated the milk and other substances to temperatures above 100°C and prevented putrefaction.[16] It was Pasteur, of course, who saw the significance of these findings to his own work on spontaneous generation.

It is my contention that all these experiments, particularly those preceding the work of Pasteur, have been stressed out of all proportion in the traditional accounts of the spontaneous generation controversy. This is partly due to the famous historical introduction to Pasteur's immortal "Mémoire sur les corpuscules organisés," in which he discussed the fermentation problem almost exclusively and which has misled historians from that time onward. This is partly due to the fact that these problems proved the most difficult to solve and continued on into the 1870s, long after other so-called examples of spontaneous generation had been explained away. This caused historians to see the whole history of spontaneous generation in terms of these last experiments on infusions, an assumption which seemed justified by Pasteur's own historical record of what had taken place.

The importance of these pre-Pasteur experiments was overshadowed by other events. In the first place, the chemists themselves believed that the ferments, although not living, were nevertheless passed through the air to initiate further changes in other media. This claim was substantiated by the lactic acid fermentation of boiled milk in contact with filtered air. In that case, since a bacterial agent was involved, the fermentation seemed to take place without any appearance of living organisms. Before Pasteur, the chemical explanation seemed to be more universal, since it could explain all fermentative phenomena, and not just the single instance of beer. It was to this particular problem that Pasteur turned in 1857, and the discovery of lactic acid "yeast" was clearly a major blow to the chemical theorists, unless, of course, the bacteria could be shown to be a result and not the cause of the fermentation.

Other events at this time added credence to the doctrine of spontaneous generation. Although Schwann's work on putrefaction and fermentation supported the opposition of the physicalists to the doctrine, his famous theory of cell development had the opposite effect. Schwann's theory of "exogeny" could be interpreted as a spontaneous generation of cells. Since "the lowest plants and animals correspond to single cells," such an interpretation was readily extended to support the spontaneous generation of organisms.

Matthias Schleiden, the German botanist, had earlier described how the nuclei or cytoblasts of new cells arose from mucus granules within a parent cell. A delicate transparent vesicle then appeared on the surface of the nucleus which expanded—eventually to produce the mature cell. This process of endogeny was denied by Schwann. "It is not the rule," he stated, "and does not occur at all with regard to many of them." Schwann ascribed the origin of most cells to a process occurring in extracellular material, the Cytoblastema, by a process very similar to endogeny in that it "takes place in a fluid, or in a structureless substance in both cases." Corpuscles first appearing in this Cytoblastema were considered by him to be either nonnucleated cells or, in most animals, the rudiments of the cell nucleus. The remainder of the nucleus was then deposited around this rudiment, and thereafter the cell membrane was formed by the deposition of new molecules:

The following admits of universal application to the formation of cells; there is, in the first instance, a *structureless* substance present, which is some times quite fluid, at others more or less gelatinous. This substance possesses within itself, in a greater or lesser measure . . . a capacity to occasion the production of cells. When this takes place the nucleus usually appears to be formed first, and then the cell around it. The formation of cells bears the same relation to organic nature that crystallization does to inorganic.[17]

In his discussion of the "theory of cells," Schwann elaborated on the metaphysical concepts behind his theory of cell development:

We set out, therefore, with the supposition that an organized body is not produced by a fundamental power which is guided in its operation by a definite idea, but is developed, according to blind laws of necessity, by powers which, like those of inorganic nature, are established by the very existence of matter. As the elementary materials of organic nature are not different from those of the inorganic kingdom, the source of the organic phenomena can only reside in another combination of these materials.[18]

Although Schwann often stated that the Cytoblastema, "lies either around or in the interior of cells *already existing*,"[19] the fact that a cell, the seat of vital activity, could be derived directly from a formless fluid which "contains only certain organic substances in solution" bore a striking resemblance to the process of heterogenesis. The influential botanist Karl Nägeli even described the process of exogeny as a spontaneous generation. When contrasting the endogenous origin of plant cells with the exogenous origin of animal cells, he remarked somewhat vaguely that in plants, "only the first cells of individuals originating through *generatio aequivoca* are formed outside cells."[20]

Even statements in support of endogeny, which seemed to support the continuity of life, were made in terms identical to heterogenetic concepts. Hugo von Mohl, who believed cells to be formed by direct division or endogenously, made the astonishing remark that, "in the regular course of vegetation this process of [free] cell-formation occurs only in the interior of cells; it may occur independently of life of the parent plant in the creation of parasitic fungi, yeast cells, etc, both in the decomposing fluid of cells and in the excreted or expressed juices."[21] In a later passage he remarked that "every globular mass wholly or partly composed of proteine compounds is capable of undertaking the function of a nucleus, clothing itself with a membrane, and thus producing a cell."[22]

I am not claiming that belief in free cell-formation necessarily implied an acceptance of spontaneous generation, but clearly the language used often implied such an association. That an anonymous supporter of Schwann's theory felt it necessary to state, "if it be true that every animal originates from an animal, and every plant from a plant, it is not true that every cell originates from a cell in those organisms,"[23] indicates that the association between spontaneous generation and free cell-formation did occur.

Some of those committed to a belief in spontaneous generation did draw support from Schwann's theory of "free cell-formation." None other than Pouchet made the connection in 1859, and the materialist Louis Büchner claimed support for his acceptance of spontaneous generation by reference to the free cell-formation of a fungus parasitic on flies:

Numerous minute free cells appear in the blood of the animal, which . . . are transformed into a microscropic fungus, *Empusa muscae*. There are various grounds for believing that these *Empusa* cells arise from a free cell-formation in the morbidly changed blood of the fly. The author entertains from his point of view, and on general grounds, no doubt as to the actual existence of a *generatio aequivoca*.[24]

Most writers who stressed the similarity between free cell-formation and spontaneous generation did so, not to substantiate a belief in spontaneous generation, but rather to bolster their opposition to Schwann's theory. This in itself indicates the low esteem in which spontaneous generation was held during this period. Robert Remak, who was the first to make the definitive statement that cells arise only from preexisting cells, remarked in 1852 that the occurrence of free cell-formation was as improbable as spontaneous generation.[25] Clearly, though, the strongest attack on spontaneous generation and free cell-formation came from Rudolf Virchow.

Virchow's famous cell dictum, *Omnis cellula e cellula,* was part of a general abhorrence of spontaneous generation, which he described at one time as "heresy or devils' work."[26] This reaction against spontaneous generation reflected his metaphysical revulsion against the physicalism of his colleagues, who, like him, had been trained in the laboratory of Johannes Müller.

The "new vitalism" which Virchow expressed in the mid-1850s did not rest on the presence of a unique vital force inherent in the substance. Rather, the new vital force "must be considered as the result of a definite joint action of physical and chemical force."[27] But any kind of life presupposed the existence of a "matrix":

Life does not reside in the fluids as such, but only in their cellular parts; it is necessary to exclude cell-free fluids from the realm of the living and intercellular material of cell-containing fluids as well. Should it be possible to present life as a whole in terms of the mechanical result of known molecular forces . . . it would not be possible to avoid designating with a special name the peculiarity of form in which the molecular forces manifest themselves. . . . Life will always remain something apart, even if we should find out that it is mechanically aroused and propagated down to the minutest detail.[28]

Other than destroying the essence of Schwann's theory, the impact of the Remak-Virchow dictum on the controversy over spontaneous generation was minimal at this time. This was partly due, of course, to the absence of certainty over the truth of the dictum. Virchow never explained the process by which cell continuity was achieved. In his famous text of 1858, *Die Cellularpathologie,* his "law of continuous development" was put forward as "an established principle," even though "absolute demonstration has not yet been afforded."[29] The main reason for its relative unimportance was that by the end of the 1850s the cell theory was changing to incorporate a theory of simple protoplasm. This concept, as will be explained in a later chapter, lent very strong support to a belief in spontaneous generation.

In the early 1840s, a fascinating controversy arose over the structure of infusorians, which again showed that spontaneous generation was a live issue. The main participants were Christian Ehrenberg on one side, and the Frenchman Felix Dujardin on the other. This debate was related to Ehrenberg's possession of a superior microscope, a fact of which Dujardin was only too well aware. By 1830 it had become possible to construct achromatic lenses, that is, lenses that were not subject to the vexation of chromatic aberration. German workers, among them Jan Purkinje and Müller and their students, began to use these new improved lenses. Included in this group was Ehrenberg. He examined fungi, molds, parasitic worms, and infusorians, all of which were seen in the 1840s as possible examples of spontaneously generated organisms. He discovered the seeds of fungi and molds and described the complex and abundant reproductive organs in parasitic worms. He also concluded that infusorians were highly complex beings having organ systems strictly comparable with the higher forms. He reported that they had a gut, stomach, excretory organs, and a heart. From these observations he concluded that "cyclical development" predominated, that given the universal complexity of all forms of life spontaneous generation was highly improbable.[30] He admitted that the spontaneous generation of infusorial eggs was still a possibility, but concluded that the general fecundity of infusorians rendered even that belief unnecessary.

Dujardin, for some unexplained reason, possessed a microscope inferior to Ehrenberg's, and admitted that, being so limited, he had no valid justification for publishing a treatise on infusorians. Despite this quandary, however, he was convinced that Ehrenberg "has yielded too easily to the rapture of his imagination." Dujardin believed, in a way very similar to Lamarck, that the organic world displayed a series of increasingly complex forms and that "the most simple infusorians are uniquely composed of a pulpy, glutinous, homogeneous substance, without visible organs but nevertheless organized."[31] This substance was the sarcode, and even in complex higher forms it was this sarcode that "appears to be the essential part." It is not surprising that Dujardin, as a Frenchman, did not conclude from this that abiogenesis was possible. Most Frenchmen were opposed to such a materialistic interpretation and had always held life to be something apart. Dujardin, however, was prepared to accept heterogenesis to account for the origins both of infusorians and parasitic worms.

The Danish physiologist, Daniel Eschricht, believed that Ehrenberg's work was mainly responsible for a change in opinion regarding the origin of parasitic worms, an assumption which can only be described as somewhat exaggerated.[32] He maintained that since the discovery of complex organs in infusorians made their spontaneous origin very doubtful, and since such organs occurred also in the parasitic worms, their origin *de novo* also became suspect. There is no evidence, however, to show that Ehrenberg's views had any impact on the worm question. The main proponents of spontaneous generation of worms had long recognized that they formed distinct species, had organs, and did reproduce by sexual means, precisely those characteristics which Ehrenberg was now showing in the infusorians.

Ehrenberg's work did, however, have an influence on the debate over the origin of infusorians, and it can be seen as one more factor leading to the general unpopularity of spontaneous generation. Naturally Dujardin's views were quoted with approval by those supporting spontaneous generation, but, since the general climate in France was opposed to spontaneous generation, his influence was minimal.

In these years the major event pertaining to the spontaneous generation debate, which overshadowed everything else, was the discovery that the origin of parasitic flukes and tapeworms could be accounted for without recourse to spontaneous generation. This came about with the discovery of parasitic life cycles, following a dramatic series of feeding experiments initiated in 1851. As explained in chapter one, such experiments were not and could not have been initiated in an attempt to solve the vexing problem; rather they occurred as a reaction to a fundamental theological issue which was open to experimentation. These experiments virtually erased any further reference to the worm problem and came as a devastating blow to the partisans of spontaneous generation. Yet, Pasteur saw fit to ignore this vital issue in his famous historical introduction! To these significant events we now turn.

One of the effects of the French Revolution was the complete reorganization of the hospitals and medical schools in Paris. One result of these changes was the birth of pathological anatomy in which anatomy was carried into pathology and diseases were considered from their effects on the tissues of the body. This resulted in detailed observations and the description of postmortem tissues.[33] In the preface to his English translation of René Laennec's *A Treatise on the Diseases of the Chest*, John Forbes quotes Dr. Clarke as reporting:

All the hospitals of Paris are under the direction of a general administration; and from the office of this board, where medical men attend during a certain part of the day to examine them, patients are sent to different hospitals. . . . Through means of this arrangement the physician of any hospital, whose attention is turned more particularily to any disease or class of diseases, by application to the central office may have such diseases sent to his own hospital. Thus a much greater number of cases of the disease . . . is brought under his observation in a given time than could otherwise, or might indeed ever, have been. To this plan we perhaps owe . . . the excellent works of Corvisant and Bayle and that of Dr. Laennec. . . . The fatal cases are generally examined after death. Dr. Forquier told me, that for twelve years that he had been physician to *La Charité,* no patient had died without being examined.[34]

These ideas quickly spread to the hospitals of London and eventually those of Germany. One result of this pathological anatomy was to increase awareness and interest in the cystic tapeworms (the Cystica of Rudolphi), and particularly "acephalocyst hydatids" (the larval stage of the tapeworm *Echinococcus*). In Britain, for example, 30 per cent of the eight hundred papers treating of parasites, and published between 1800 and 1860, dealt with these human hydatid infections.[35] Many and detailed descriptions of these cysts appeared, but their origin was clouded in mystery. Some scientists, like the English pathologist Richard Bright, considered them to be "independent animals, existing without any vascular connection with the body in which they are developed," of the origin of which "we are so completely ignorant, that it would be vain to hazard a conjecture on the subject."[36] Others, I think the majority, considered them to be products of diseased body tissues:

Humidity, abundance and the bad or vegetable quality of the nourishment of an animal, are unequivocal means of producing acephalocysts . . . irritation in fact, of a specific kind, has been excited by which a state favourable to their development has been produced. . . . That irritation must be sufficient to cause the exhalation of a particle of lymph, that lymph . . . becomes organized, acquires step by step an individual existence, it will be the minimum of organization of independent vitality, but still, when its separation is achieved, it will be a living being.[37]

Even the physiologist Allen Thomson, who was always very reluctant to accept the possibility of spontaneous generation, had to admit in 1840 that "these facts appear to us to speak strongly in favour of the occasional occurrence of spontaneous generation."[38]

One cannot overemphasize that these cysts did seem to provide a perfect example of a spontaneously generated being. They carried no reproductive organs and were located in parts of the body with no outlet to the exterior. In addition, specific cysts were restricted to specific parts of the body, supporting the idea that they were characteristic morbid products of specific organs.

John Goodsir, the famous Scottish anatomist who was a firm believer in cell and life continuity, managed to retain such beliefs in the face of the hydatid problem. While admitting that the location of hydatids deep within the tissues prevented their propagation, he managed to deny their spontaneous origin by reference to the benevolence of nature. In a passage that reflects a blend of Scottish stoicism and British natural theology, he remarked: "As the hydatid is by no means of unfrequent occurrence in the liver and other internal organs, this limitation of the increase appears to be a beneficent law of nature, for the purpose of preventing the fatal termination which the rapid increase of these animals would infallibly produce."[39]

The situation regarding the other parasitic worms, the trematodes, was one of complete and utter confusion. In 1831 Karl Mehlis described a small infusorian (miracidia) hatching from the eggs of a trematode adult, a result which must have been received with complete incredulity. The existence of *"Königsgelben Würmen"* in the digestive glands of snails (i.e., sporocyst-rediae generations) was known, and since they too lacked reproductive organs, they seemed to provide another example of spontaneously generated beings. It was also known that these worms produced masses and masses of tailed individuals which swam about in the water and were considered to be typical infusorians of the genus *Cercaria*. It is amusing to note that Richard Owen, noting the superficial similarity between cercariae and sperm, considered the latter to be parasites of man and named them *Cercaria hominis*! There was no evidence to suggest that these cercariae metamorphosed back into the rediae from which they came, and indeed Carl Theodor von Siebold had described their encystment (metacercariae) in a sort of chrysalis without having any idea of their future development. Von Siebold had further confused matters by reporting the enclosure of *Königsgelben Würmen*-like bodies in an infusorian animal produced by the hatching of the eggs of the trematode *Monistomum mutabile*. In 1847 G. Gros described in detail the production of tapeworms from the digestive gland of *Sepia*. This was the last published report describing the production of a parasitic worm by spontaneous generation.[40]

Rudolf Leuckart, speaking specifically to the worm problem, pointed out that new facts would have altered little "had not a change in the direction and method of biological study given a fresh impulse to helminthology."[41] This fresh impulse came from the publication in 1842 of Japetus Steenstrup's *On the Alternation of Generations*, translated into German in the same year, and into English in 1845. Carl von Siebold clearly pointed out the impact of this work when he wrote:

By degrees a mass of observations accumulated and constituted a complete chaos of seemingly irregular phenomena which broke down every barrier hitherto set up by the acknowledged laws of animal existence and propagation, until the penetration of the Danish naturalist, Steenstrup, succeeded in evolving a certain order out of this confusion, by the discovery therein of a hidden, underlying law of nature.[42]

Steenstrup drew together all the disjointed facts then known about trematodes, added a few observations of his own, and explained them by reference to the concept of the alternation of generations. This concept, which was drawn up primarily to explain coelenterates' life cycles (with the alternation of polyp and medusa forms), is most fascinating in that he defined it in terms which contradicted the very foundations of the viewpoint of those opposed to spontaneous generation, namely, that like begets like. The concept was defined by Steenstup thus:

An animal producing an offspring, which at no time resembles its parent, but which, on the other hand, itself brings forth a progeny, which returns in its form and nature to the parent animal, so that the maternal animal does not meet with its resemblance in its own brood, but in its descendants of the second, third or fourth degree of generation; and this always takes place in the different animals which exhibit the phenomenon, in a *determinate* generation, or with the intervention of a *determinate* number of generations.[43]

Steenstrup followed the development of *Cercaria echinata,* released from *Königsgelben Würmen* in the snail, which von Siebold had shown previously to encyst in snails. He extracted the animal from the cysts, after various periods of encystment, and found that they formed typical flukelike forms. Steenstrup, by a magnificent piece of intuition, drew these data together to outline in general detail a typical fluke life cycle: "*Cercariae* are the *larvae* of *Distomata,* and that they become true *pupae* [metacercariae]; that the animals in which the *Cercariae* originate and grow up are peculiar *trematode* individuals [sporocysts and rediae], descended in the second and third generations from *Distomata,* and consequently not produced directly from their ova."[44] The progeny of trematodes owed their existence to the "parent-nurse" (sporocyst) and "nurse" (daughter sporocysts and rediae) generations within the snail digestive gland, "which in external form and partly in internal organization, differ from the animals into which their progeny is afterwards developed."[45] Steenstrup assumed that the distome adult became mature in the snail host in which it was encysted, rather than in a vertebrate host to which it had to be transferred.

In many respects the events in parasitology over the next two decades were determined by the reaction of von Siebold, by this time an established and famous figure among German academics, to the work of Steenstrup. He reported on Steenstrup's work the same year it appeared and was obviously much impressed by it. He pointed out in his article that it would be unlikely that the metacercariae would transfer into an adult worm without a further change of host. Von Siebold came to realize that many encysted cercariae were located in organisms which could not be consumed by the organism in which the adult fluke would later develop. These parasites, doomed to perish, were considerd by him to be "strays."[46] Dujardin, in France, denied that metacercariae could ever become adult worms; while a sporocyst could become a distome adult, he did not "believe this to be by shutting itself in this excreted prison."[47] It was this concept of "strays" which led to the initiation of feeding experiments, for, unlike the broader question of spontaneous generation, it was capable of refutation by experimental means.

There is evidence to show that Dujardin and von Siebold came to hold the stray-concept from a belief in or sympathy with the doctrine of spontaneous generation. Dujardin certainly accepted that parasitic worms could arise spontaneously. Although the same cannot be said for von Siebold, he may well have accepted spontaneous generation in his early career, for he was a pupil of Carl Rudolphi, and had written an article on worm development in *Die Physiologie* of Carl Burdach, a well-known supporter of spontaneous generation.

The concept of "strays," when applied to trematode life cycles, was of minor importance because it did not fundamentally alter the scheme set up by Steenstrup. Von Siebold and Dujardin, however, carried the concept over into the tapeworms by assuming that the cysticerci, hydatids, and coenuri (Rudolphi's Cystica) were "nothing other than strayed tapeworms which remain immature and undergo dropsical degeneration."[48] The Cystica were thus regarded as a pathological condition and not part of a normal life cycle; in the words of Dujardin, they were "une sorte de monstruosité."[49] This assumption was of major significance to the overall picture of cestode life cycles, and it was this assumption that Küchenmeister found so distasteful.

Friedrich Küchenmeister at this time was an unknown physician living in Zittau, and he was only thirty years old when he burst upon the scene. His biographer paints the picture of a very religious man, his study being described as more befitting a theologian than a physician, with holy books lying open on chairs and tables.[50] Küchenmeister wrote books on Luther and biblical zoology, among other subjects, and was obviously a fanatic regarding cremation. He conducted the

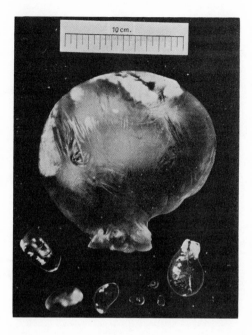

The larval cysticercus stage of *Taenia serialis* from the body cavity of a rabbit. This so-called "coenurus" cyst buds off smaller secondary cysts which are seen below the primary cyst. From an article by A. G. Hamilton, "The occurrence and morphology of *Coenurus serialis* in rabbits," in *Parasitology* 40 (1949), reproduced here with the permission of Cambridge University Press.

first four cremations in the German states, was founder of the Dresden Cremation Club (which held the ominous name of *Die Urne*), and wrote on burial rites in the Bible. With regard to parasites, Küchenmeister took a very simple stand, which rested on his teleological view of nature:

If the caudal bladder [Schwanzblase], named by von Siebold, were a diseased structure, it would be contrary to the wise arrangement of Nature which undertakes nothing without a purpose.... The larva [Cysticercus] is not a wandering strayed dropsical tapeworm nurse, in the sense of Steenstrup's Alternation of Generations, but a tapeworm larva provided with a temporary organ [Schwanzblase], probably functioning as a reservoir of nourishment.[51]

Küchenmeister took this disagreement with von Siebold and other members of the "biological establishment" far beyond the mere details of tapeworm life cycles. He regarded it as his scientific duty not simply to describe facts, but to bring the Law of God into recognition. "Such a theory of error," he wrote, referring to the stray-concept of von Siebold, "contradicts the wisdom of the Creator and the laws of harmony and simplicity put into Nature."[52] His writings are imbued with a fervor and intensity that one little expects in writings on parasites, and it is not surprising that the debate between him and von Siebold involved a great deal of invective: "Therefore it is our duty, to analyze this whole theme with all the strength of our criticism and without regard to our finer emotions. We must uncover the blemishes of the previous theory and persecute the errors step by step."[53]

Küchenmeister was, of course, correct; the cysticerci are a necessary part of the tapeworm life cycle. Most of the evidence at that time, however, pointed in the opposite direction. He was not only taking issue with some of the great names of that period—such as von Siebold and Dujardin—but also with the empirical evidence at their disposal. To do this obviously required a deep commitment, the type of commitment that comes from deeply held religious views.

The factual pieces of evidence with which Küchenmeister took issue were many. In the first place, there was a great interest at that time in the encysted hydatids and *Trichinella* (a nematode parasite) of man, to which I have referred above. These organisms were obviously "strays," trapped within man's muscles and never able to reach their final host, since man is at the end of a food chain.

Another line of reasoning which led to the stray-concept begins in the work of Steenstrup. Steenstrup was concerned primarily with the trematodes, but in a very short section of his work he posed the question of whether tapeworms display a similar alternation of generations. This was the very question with which the parasitologists of the next few years concerned themselves. This question was intimately related to the problem of the nature of the tapeworm body, which had long proven a puzzle. The body of a tapeworm is divided into a series of compartments, or proglottids, each one of which contains a complete set of reproductive organs. Some regarded each of these proglottids as equivalent to an individual organism and thus viewed the tapeworm as a polyzoic form. Others viewed the complete tapeworm as the individual and each proglottid as a segment of a monozoic worm, exactly equivalent to the segments of annelids and arthropods.

Pierre-Joseph van Beneden, in his *Les Vers cestoides* of 1850, saw Steenstrup's work as playing a fundamental role in this very question and indeed mentioned him only in this context. Steenstrup and van Beneden viewed the tapeworm as a polyzoic organism, and thus regarded the individual proglottids as being equivalent to the trematode adult, and the scolex, from which these proglottids bud off, as being equivalent to the sporocyst "nurse" generations of the trematode. Tapeworms, in other words, show an "alternation of generations" almost identical to that of the trematodes, but one in which the budding takes place externally. "We learn that this internal worm of the monostome [sporocyst] gives birth to an endogenous generation whose individuals have the form of cercariae, and that the latter develop into distomes which alone are sexually mature. In the cestodes, the same phenomenon takes place, with this single difference, that the budding takes place to the exterior rather than to the interior of the scolex."[54] In this interpretation, which sees a fairly exact correspondence between cestodes and trematodes, it is not surprising that the cystic stages came to be regarded as superfluous since there are no similar stages in the trematode cycle.

Finally, and I think the most important argument as far as von Siebold was concerned, there was the obvious fact that the cysticerci appeared to be degenerate in comparison with other known tapeworm larvae. Those of the aquatic pseudophyllids, the plerocercoid larvae, show a superficial resemblance to tapeworms and, indeed, some of them even show surface segmentation.[55] He was also aware of the rather unusual structure of *Cysticercus fasciolaris* from rat livers, in which the scolex is everted, bears a longish neck, and has a small bladder attached at the posterior end. This form, with its tapewormlike appearance, was seen to be capable of transforming into an adult worm when eaten by a cat: "I have good reason to believe, that with the exception of *Cysticercus fasciolaris* . . . no other single degenerated cestode bladderworms can return from their dropsical state to their former state, so as to become suitable for the production of sexual individuals."[56]

Von Siebold and Küchenmeister approached this problem from very different points of view. Von Siebold saw his task as determining which cysticercus degenerated from which tapeworm species, a problem that could be solved only by microscopical examination of their respective structures. Küchenmeister, on the other hand, was interested in showing that these cysticerci could develop into adult worms, a problem that could be solved only by feeding cysts to various animals. The era of experimental parasitology was born when, between March 18 and April 9 of 1851, Küchenmeister fed foxes with a typical cysticercus, *Cysticercus pisiformis,* obtained from rabbits.[57] This experiment initiated an explosion in feeding experiments over the next decade or so.

In the period which followed, von Siebold and Küchenmeister differed in their interpretation of the experimental results. Von Siebold maintained that the different types of cysticerci were all degenerated forms of the same species, their forms being determined by the host into which they had strayed, but Küchenmeister held that each type of cysticercus was a larval form of a distinct species. The issue, however, was complicated by additional factors. In the first place there was a great deal of personal antagonism between von Siebold and Küchenmeister. Von Siebold was a highly respected and powerful holder of a full professorship at Munich, a position which carried with it prestige and power.[58] In contrast, Küchenmeister was an unknown figure without the thorough training in biological techniques that so characterized nineteenth-century German universities. As a result Küchenmeister had great difficulty in identifying the species of tapeworms he obtained in his feeding experiments, a fact that von Siebold was only too delighted to point out. Identification of the species obtained was of course the major issue in the controversy, since von Siebold claimed that the different types of cysticerci developed into the same species of tapeworm.

Poor Küchenmeister had a disastrous beginning. In his initial experiment he reported that *Cysticercus pisiformis* developed into *Taenia crassiceps*, but a few weeks later he corrected this to identify the tapeworm as *T. serrata*. The following year he changed his mind again and erected an entirely new species, *Taenia pisiformis*.[59] Von Siebold triumphantly exclaimed:

This rapid succession of contradictory statements would naturally have the effect of disinclining the medical public to believe in the possibility of the transformation of the *Cysticercus* into a *Taenia*. But even with helminthologists themselves Küchenmeister's statements could not meet with any real acceptance, as it was but too apparent from the whole exposition of his researches, that he was still, as he himself confesses, deficient in helminthological information.[60]

In 1852 and 1853 von Siebold fed four types of cysticerci to dogs, *C. pisiformis, C. tenuicollis, C. cellulosa,* and *Coenurus cerebralis,* and in each case claimed to have obtained the same species of adult tapeworm, *Taenia serrata*.[61] From these experiments he concluded that "all these cystic worms are only the degenerated embryos and scolices of a single species of *Taenia.*"

The results of these feeding experiments... contradict the belief that the cysts of these worms have a physiological, and not a pathological significance. For all the Cystica mentioned are produced from a single species of tapeworm, namely the *Taenia serrata,* and it only depends upon the nature of the spot to which these embryos have been transplanted after having completed their immigration, whether they degenerate into *Coenurus cerebralis, Cysticercus pisiformis* or *tenuicollis,* etc.[62]

Von Siebold urged helminthologists not only to feed cysticerci to the definitive host, but also to feed tapeworm eggs to animals in order to examine the development of the bladderlike stages. During the next few years this is exactly what occurred, and experimenters such as Küchenmeister, van Beneden, and Leuckart carried out numerous two-way feeding experiments, cysticerci to adult and eggs to cysticerci. By 1855 they had concluded that each cysticercus developed into a specific tapeworm and Leuckart was able to claim,

It is by the results of my experiments that I have become convinced that *Taenia solium, T. serrata, T. cysticercus tenuicollis* and *T. coenurus* are by no means identical, and are not variations of a single species as a well respected authority has lately believed. The eggs of *T. solium* never produce the coenurus. . . . There is no longer any authorization to call cysticerci hydropsical worms. . . . The cysticerci are an absolutely normal phase of development.[63]

The episode ended in a rather unfortunate fashion. In 1852, the French Academy set a prize essay on the problem of the development and transmission of intestinal worms. The prize was carried off in 1853 by P. J. van Beneden, but his winning entry, "Mémoire sur les vers intestinaux," was not published until 1861.[64] Küchenmeister received only an honorary mention. Van Beneden's prize may well have been merited, since the essay is an excellent résumé of the new discoveries in parasitology which forever removed these worms from the realm of spontaneously generated animals. However, there is also the rather uncomfortable feeling that the prize went to the man who occupied a powerful chair at the University of Louvain. Moreover, van Beneden made two claims on his own behalf that were certainly unjustified. He claimed the honor of having put forward the concept of worm transmigrations first, before anybody else had ever "dreamed" of such a thing, and also that it was he and not Steenstrup who had developed the true concept of the alternation of generations as an alternation of reproductive methods and not merely of body form.[65] Clearly, however, the honor of having refuted the old beliefs regarding parasitic worms rests with Steenstrup and Küchenmeister; it was their concepts that led the other parasitologists to discover the various lifecycles and thereby to present a model which completely replaced the old idea that worms were generated from diseased host tissue.

Nematodes, or round worms, had rarely been regarded as parasites which were spontaneously generated. They are an extremely common and ubiquitous group of organisms, about half of which are freeliving and half plant or animal parasites. They are characterized by a truly remarkable constancy of body form which renders their identification a frustrating and tedious task. It was precisely this ubiquity and constancy however which led to the assumption that they could be picked up from external sources rather than produced spontaneously from diseased host tissue. The initiation of feeding experiments with nematodes by Ernst Herbst in 1845 and again in 1851 had little impact on the problem of spontaneous generation.[66]

The events surrounding the discovery of tapeworm life cycles in the first half of the 1850s did, however, lead to the assumption by Küchenmeister that the human encysted *Trichinella spiralis* might well be a larval stage of another nematode. *Trichinella* had been used by von Siebold to justify his stray-concept, since clearly any encysted worm in man could not be a part of a normal life cycle. Küchenmeister's belief led Leuckart and Virchow to carry out feeding experiments with the organism. In 1859 Virchow fed living *Trichinella* from a dead patient to a dog and almost immediately obtained mature, sexually ripe nematode adults. One of the difficulties was, however, that both Leuckart and Virchow were expecting a radical metamorphosis to take place as in the cestode experiments. As Edward Reinhard pointed out, the discovery that, *"Trichinella spiralis* in the intestine of the dog, becomes sexually mature in a very short time *but does not undergo any further metamorphosis,"* came as a "thunderclap."[67]

Virchow's involvement with the work on parasitic life cycles in the 1850s provided him with strong empirical evidence to support his metaphysical abhorrence of spontaneous generation. It is not surprising to see that his famous cell dictum of 1858 was preceded by reference to parasitic worms:

At the present time, neither fibres, nor globules, nor elementary granules, can be looked upon as histological starting-points. As long as living elements were conceived to be produced out of parts previously destitute of shape, such as formative fluids, or matters, any one of the above views could of course be entertained, but it is in this very particular that the revolution which the last few years have brought with them has been the most marked. Even in pathology, we can now go so far as to establish, as a general principle, *that no development of any kind begins* de novo, *and consequently as to reject the theory of equivocal* [spontaneous] *generation just as much in the history of the development of individual parts as we do in that of entire organisms.* Just as little as we can now admit that a taenia can arise out of saburral mucus, or that of the residue of the decomposition of animal or vegetable matter an infusorial animalcule, a fungus, or an alga, can be formed, equally little are we disposed to concede . . . that a new cell can build itself up out of any non-cellular substance. Where a cell arises, there a cell must have previously existed (*omnis cellula e cellula*).[68]

Clearly the discovery of parasitic lifecycles and increasing criticism of Schwann's theory of cell development were important factors in the decline of the spontaneous generation concept in these years. It forced partisans of spontaneous generation to rely solely on the unexplained origin of microscopical forms, but even here experimental evidence was undercutting their beliefs.

In the 1850s the rising school of German materialists, led by Louis Büchner, Carl Vogt, and Jacob Moleschott, introduced into the debate another element which was to play a very significant role in the post-Darwin years. Atheists and outspoken proponents of social revolution, and committed to an a priori belief in the possibility of abiogenesis, they turned their attention to the origin of the first organisms on the surface of the earth. As Büchner argued in his infamous *Kraft und Stoff,* which by 1904 had been translated into fifteen languages and gone through twenty-one German editions:

The general law at present seems to be *omne vivum ex ovo,* i.e., all living beings derive their origin from existing germs produced from homogeneous parents, or by direct generation, as from an egg, a seed, or by division, budding etc.... Those who are not satisfied with the traditions of the Bible, necessarily raise the question *whence?* and *how?* i.e., the question of the first origin of organic beings. If every organism is produced by parents, whence came the parents? Could they have arisen from the merely accidental or necessary concurrence of external circumstances and conditions, or were they created by an external power? And if the first supposition be true, why does it not happen today?[69]

From the basic assumption "that there are no other forces in nature beside the physical, chemical and mechanical," the materialist had to "infer irresistibly that the organisms must also have been produced by these forces."[70] Obviously they also had to infer that a continuous spontaneous generation of simple organisms was possible if the conditions were favorable. In the 1850s Bücher could substantiate such beliefs by claiming, with justification, that "the question of a *generatio aequivoca,* or spontaneous generation, is not yet settled."[71]

Bücher attempted to draw support for his belief in spontaneous generation from the work of the chemists, who by the mid-nineteenth century were agreed that organic substances were constituted of the same elements as inorganic matter, although at a higher level of complexity. Many chemists, including John Berzelius himself, assumed that this higher level of complexity required the existence of a nonphysicochemical force. The German physiologist, Johannes Müller, in 1835, wrote that, "the mere accidental coming together of elementary components is not capable of producing organic matter," and thus denied the possibility of abiogenesis. But his vitalistic assumptions allowed him to believe in the feasibility of heterogenesis: "All the organic matter which is spread over the surface of the earth, originates in living beings. Death... annihilates the individual, while the organic matter which formed this individual, while it is not reduced to binary compounds, is still capable of receiving new life, or in other words, of nourishing other living bodies."[72]

The work of Justus von Liebig, and the numerous syntheses of organic compounds in the laboratory, including that of urea by his friend Friedrich Wöhler in 1828, did banish "vital forces" from the realm of organic chemistry. Contrary to popular opinion, however, these events did not lead chemists to deny the occurrence of vital forces in living organisms. In a recent paper, John Brooke pointed out that it is, "the common failure to distinguish two separate issues, namely the peculiarity of a compound and the peculiarity of an organism, which is largely responsible for perpetuating the popular appraisal of Wöhler's work."[73] Despite their acceptance of the basic similarity of organic and inorganic matter, most chemists were only too conscious of the differences between their organic substances and the living organisms which produced them. Many of them, such as the great Liebig, while denying "vital forces" in organic chemistry, still applied them to physiological phenomena within an organism. As Timothy Lipman has pointed out, the importance of the vital forces in Liebig's physiology was not the force per se, but the laws of vitality which were in harmony with laws pertaining to inorganic forces. "For Liebig it was as valid to attribute the laws and effects of vitality to a vital force as it was to attribute the laws and effects of gravity to a gravitational force."[74]

Holding to such beliefs, Liebig did not view his own work or Wöhler's synthesis as in any way justifying a belief in the possibility of spontaneous generation. Indeed, Liebig explicitly denied any possibility of its occurrence: "If anyone assured us that the palace of the king, with its entire internal arrangement of statues and pictures, started into existence by an accidental effort of a natural force, which caused the elements to group themselves into the form of a house . . . we should meet such an assertion with a smile of contempt, for we know how a house is made."[75]

Similarly, in a living organism, he argued that "our reason tells us that the idea must have had an author." The first living organism was created by God and all living organisms originate only from pre-existing parents.

Bücher, however, saw the unity of organic and inorganic in different terms. In a passionate attack on Liebig's use of vital forces, he accused Liebig of being an "amateur and promenader" when it came to physiology and asserted: "It is as clear as the sun: the same elementary materials enter the organism as in inorganic bodies, and no real naturalist doubts that forces are but qualities or motions of matter, and that, consequently, no other forces can act in organic beings but such as pertain to . . . the general forces of nature." He pointed to the synthesis of organic compounds as "striking proof" against the view that "chemistry could not imitate the products of the organism," and added, "we know now how far we may yet advance."[76] To Büchner, the synthesis of organic compounds appeared as an initial stage leading to the eventual spontaneous generation of organisms, the distinction between organic compounds and simple living organisms being meaningless to him. Such an interpretation of the new organic chemistry received strong support a few years later with the appearance of *The Origin of Species*. Once a universal evolution of organisms was accepted, then clearly the synthesis of organic compounds could be viewed in Büchner's terms.

Curiously, Büchner's fellow materialist, Carl Vogt, initially denied the occurrence of spontaneous generation on the basis of its association with *Naturphilosophie*. In 1851, and again in 1858, he produced a German translation of Robert Chamber's *Vestiges of the Natural History of Creation,* inserting notes to correct the author's support for spontaneous generation.[77] A year later, however, he questioned the validity of Schwann's experiments and argued that the passage of air over chemicals and the heating of air changes its nature in some way.[78] But in 1863 he once again expressed great doubt over the possibility of producing simple organisms spontaneously. "In spite of all contrary statements," he wrote, "the spontaneous generation of organisms remains beyond the reach of observation and experiments."[79]

However, Vogt's belief in a series of universal catastrophes to account for the history of the earth's crust allowed him to accept the occurrence of abiogenesis to stock each successive geological stratum. Like Büchner, he drew attention to the synthesis of organic compounds, but admitted that the production of organisms was a different matter. He did not rule out the possiblity, however, of eventually producing living organisms in the laboratory.

The continuing hostility of British scientists to spontaneous generation was expressed in their attacks on Chamber's *Vestiges of Creation,* which appeared in 1844. In this work, an "organic creation by law" had been postulated. He claimed support for his acceptance of spontaneous generation by reference to the synthesis of urea and allantoin, which he saw as a realization of "the first step in organization." Not only did he claim support for the doctrine from parasitic worms, but he also made reference to the appearance of ice crystals, "precisely resembling a shrub."[80] (Unfortunately, the modern addiction to central overheating has blotted out this remarkable handiwork of "Jack Frost," once so delightful a part of childhood.)

Two years later Chambers again stressed that it was inconceivable that the creation should not come under the rule of universal laws:

I am at the very first struck by the great *a priori* unlikelihood that there can have been two modes of Divine working in the history of Nature—namely, a system of fixed order or law in the formation of globes, and a system in any degree different in the peopling of these globes with plants and animals. Laws govern both: we are left no room to doubt that laws were the immediate means of making the first; is it to be readily admitted that laws did not preside at the creation of the second also?[81]

That such laws could indeed create life was substantiated, according to Chambers, by the experiments of one H. W. Weekes, Esquire, who, in a series of controlled experiments lasting three years and three months (1842–1845), obtained living acari (ticks and mites) in a solution of potassium ferrocyanate subject to "electrical influence."[82] Such carefully contrived and rigorous experiments could not, argued Chambers, be set aside "merely with scoffs and jests." The British were not impressed; to them spontaneous generation seemed to involve "contrivance, without a contriver and design without a designer," in opposition to the familiar lines of William Paley, whose book had thrilled and charmed Charles Darwin:

There cannot be design without a designer; contrivance without a contriver; order without choice; arrangement, without anything capable of arranging; subserviency and relation to a purpose, without that which could intend a purpose; means suitable to an end, and executing their office in accomplishing that end, without the end ever having been contemplated, or the means accommodated to it.[83]

Despite the optimism of the materialists, the doctrine of spontaneous generation was in disarray by 1859. Yet, curiously, in the next decade it was once more to achieve a new popularity in Germany and even in Britain.

FIVE

POPULARITY RESTORED

THE SEVENTH DECADE IN GERMANY
AND BRITAIN

That Pasteur's work on spontaneous generation appeared when the doctrine was on the wane, lends support to the popular notion that the issue was finally resolved by him in the 1860s. This was true only in France, however. In Britain and Germany the theory of spontaneous generation held a degree of support that it had not enjoyed since the early years of the nineteenth century.

Two major reasons for this remarkable recovery stand out. First, Charles Darwin's *Origin of Species* appeared in 1859, and, although Darwin and Thomas Henry Huxley publicly stated that life could have been originally created, the French and Germans realized that any theory of evolution based on natural causes demanded an abiogenetic origin of life. But while the materialists saw this implication of Darwinism as a validation of their metaphysical position, most biologists were only too well aware of the predicament. They were almost forced to accede to a belief in abiogenesis at the same time as experimental findings seemed to have exploded the possibility of heterogenesis. Hence came the great dilemma: if simple living organisms could not be generated from a rich organic "soup," it was hardly possible that they could arise from solutions of simple inorganic substances.

Secondly, the cell theory was modified in the 1860s so that protoplasm came to be considered the structural unit of life. Moreover, this protoplasm was viewed as an essentially simple substance, theoretically capable of being produced directly from inorganic matter. That the simplest organisms were generally regarded merely as naked lumps of protoplasm added credence to the belief that they, too, could be produced spontaneously. In addition to these factors, the assumed nature of cholera and choleralike diseases allowed for a *de novo* origin of disease. Proponents of spontaneous generation saw such disease theories as a vindication of their position.

The materialists were quick to see the implications of Darwin's theory. Having argued earlier in support of spontaneous generation, they now saw such views triumphantly vindicated. The aspect of Darwinism to which they responded was not so much the mechanism of natural selection as the theory of descent, which they saw unifying the organic and inorganic realms into one materialistic world system. In his preface to the 1864 English edition of *Kraft und Stoff*, Louis Büchner remarked: "I could not know [in 1854] that the dogmata concerning the non-existence of primeval (spontaneous) generation, and the immutability of species, which were then considered almost too sacred for attack, would soon experience such severe shocks and that the celebrated theory of Darwin would reduce the whole organic world, past and present, to one great fundamental conception."[1]

Acceptance of abiogenesis was naturally a cornerstone of materialist philosophy. In the eyes of the materialists, to deny its occurrence was to widen the gap between life and nonlife and to imply the existence of vital forces. As Büchner remarked in 1856: "The first origin of organic beings upon the earth contains, in fact, the gist of the whole matter in dispute in regard to vital force."[2] Commitment to materialism and the unified world view was such that any arguments over the present occurrence of spontaneous generation were of little interest to the materialists. Nevertheless, in the 1860s they were even more justified in maintaining the question of spontaneous generation to be "still an open one." Similarly, Büchner's statement of 1855 now seemed vindicated even more:

Although modern investigations tend to show that this kind of generation, to which formerly was ascribed an extended sphere of action, does not exactly possess a scientific basis, it is still not improbable that it exists now, in the production of minute and imperfect organisms. . . . Spontaneous generation exists, and higher forms have gradually and slowly become developed from previously existing lower forms, always determined by the state of the earth, but without immediate influence of a higher power.[3]

Justification for the statement that the question of spontaneous generation was "still an open one" arose from changing concepts in cell theory. Attention moved away from the cell wall to the cell contents. In 1860, Max Schultze, the German embryologist, redefined the cell in these terms:

a cell is a lump of protoplasm inside of which lies a nucleus. The nucleus, and the protoplasm as well, are division products of like constituents of another cell. This must be added in order to adhere to the concepts of nucleus and cell as distinct from other structures which perhaps look similar.

The cell leads, so to speak, a life of its own, the vehicle of which is again primarily the *protoplasm,* although the nucleus also definitely plays a significant role, but one which cannot yet be described in detail.

One must *emphatically* keep in mind, first of all, that the protoplasm is closed off on the outside by nothing but *its own peculiar consistency* differing from that of the surrounding watery fluid. . . . Thus, the cell *without a membrane*, may remain to a certain extent independent of outside influences.[4]

To Schultze, the nucleus was a necessary part of the cell, but its role was unknown. Since it was seen by him to be a special kind of protoplasm, it is not too surprising to see even the nucleus becoming an unnecessary aspect of the cell. It was Ernst Haeckel who was finally responsible for seeing the cell and primitive life as "a simple little lump of albuminous combination of carbon."

Haeckel's views on the nature of protoplasm formed part of the broader issue of evolution to which we shall return later. Here it suffices to say that Haeckel believed the most simple life to be cytods, undifferentiated organisms without a nucleus, which were "similar to inorganic crystals, in being homogeneously composed of one single substance."[5] From these primitive beings more complex ones with nuclei and cell membranes could then arise. The point was, however, that these simple organisms could arise by spontaneous generation, by heterogenesis or abiogenesis. Since the membrane-bound cells of higher forms were seen to be composed of exactly the same simple substance as these cytods, Haeckel's work lent strong support to the protoplasmic theory.

In Britain the protoplasmic theory received popular support also, so much so that, by 1873, John Cleland could remark: "In the present day the protoplasmic element has assumed an enormous importance, casting the nucleus into the shade, while the reign of cell walls has come to an end altogether."[6] Even that opponent of spontaneous generation, John Drysdale, had to admit, as late as 1874, that "the property of vitality is restructed solely to a universally-diffused, pulpy, structureless matter."[7]

In 1868, Thomas Henry Huxley presented his famous address, "On the Physical Basis of Life," in which he concluded that "protoplasm, simple or nucleated, is the formal basis of all life. It is the clay of the potter; which, bake it and paint it as he will, remains clay, separated by artifice, and not by nature, from the commonest brick or sundried clod."[8] Although in some ways this paper represented a change in thinking from his earlier paper of 1853, "The Cell Theory," in which he had stressed the importance of the cell wall, he was fundamentally putting forward the same mechanistic concepts that "life is the cause and not the consequence of organization" and that "the vital forces are molecular forces."[9]

Huxley was opposed to the concept of Schwann that vital activity results from the organization of matter into cells. On the contrary,

Huxley believed as firmly in 1868 as he did in 1853 that, "The vital phenomena are not necessarily preceded by organization nor are in any way the result or effect of formed parts, but that the faculty of manifesting them resides in the matter of which living bodies are composed."[10] That life is possible without cells, and that the appearance of cells is a manifestation of molecular forces inherent in matter, received added stimulus from the discovery of *Bathybius haeckelii*.

In 1857 Huxley had examined some mud dredged from the sea bottom during the voyage of H.M.S. *Cyclops* and reported the presence of "a multitude of very curious rounded bodies,"[11] which he surmised could not be organic. In 1868 he reexamined this same mud under a more powerful microscope, and, no doubt influenced by the work of Haeckel, reported that it contained "innumerable lumps of a transparent, gelatinous substance" in which were embedded "granules, coccoliths and foreign bodies."

I conceive that the granule-heaps and the transparent gelatinous matter in which they are embedded represent masses of protoplasm. Take away the cysts... [it would] very nearly resemble one of the masses of this deep sea "Urschleim", which must, I think, be regarded as a new form of those simple

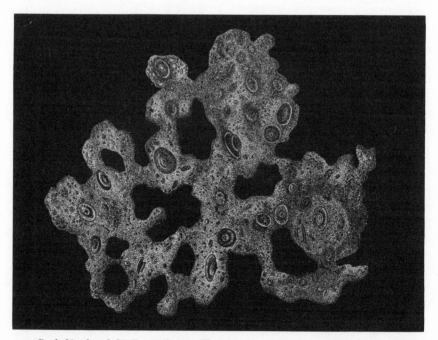

Bathybius haeckelii. From C. Wyville Thomson, *The Depths of the Sea* (London, 1873).

animated beings ... described by Haeckel. ... I propose to confer upon this new "Moner" the generic name *Bathybius* and to call it after the eminent professor of zoology in the University of Jena, *B. Haeckelii.* [12]

A year later Haeckel's "Monographie der Moneren" appeared in an English translation.[13] Thus the notion that the sea bottom was covered with a layer of protoplasmic *Urschleim,* vindicated by the discovery of *Bathybius,* came to have widespread appeal in Britain.

Haeckel saw the discovery of *Bathybius* as a triumphant vindication of his acceptance of abiogenesis, while even Huxley, having grown up in an environment in which the theological repercussions of a belief in abiogenesis were far more apparent, was not prepared to deny its possibilities. In his 1870 address to the British Association for the Advancement of Science, Huxley remarked, "I must carefully guard myself against the supposition that I intend to suggest that no such thing as abiogenesis has ever taken place in the past or ever will take place in the future. ... All I feel justified in affirming is, that I see no reason for believing that the feat has been performed yet."[14] John Browning in 1869 pointed out that it was no longer largely accepted that life was indeed transmitted only from one organism to another. "There is no boundary line between organic and inorganic substances. ... Reasoning by analogy, I believe that we shall before long find it an equally difficult task to draw a distinction between the lowest forms of living matter and dead matter."[15]

In many ways, Haeckel's response to the problem of abiogenesis was similar to that of the dogmatic materialists. He built up a complete *Weltanschauung* around Darwinism, seen as the doctrine of descent, in which an abiogenetic origin of life was an a priori necessity. He clearly distinguished between heterogenesis and abiogenesis and avoided at least one pitfall—he never attempted to justify a belief in abiogenesis by reference to work on heterogenesis, even at the time when that latter doctrine was not fully discredited.

In his first publication of 1862, *Die Radiolarien,* Haeckel had quietly accepted the transmutability of species and constructed a genealogical tree of the radiolarians. It was in the form of a working hypothesis, but even at this stage he remarked: "The greatest fault of the Darwinian theory is that it yields no clues on the origin of the primitive organism from which all the others have gradually descended—very probably a simple cell. When Darwin assumes a special creative act for this first species, he is not consistent at any rate and, I think, it is not intended to be taken seriously."[16]

A little later, he progressed far beyond this timid acceptance of Darwinism. This occurred at the Scientific Congress of 1863 and particularly in his great masterpiece of 1866, *Generelle Morphologie der Organismen*. All organisms, including man himself, were seen to have evolved from a few simple organisms, which in turn had arisen from inorganic matter. He saw in the work of Darwin a vindication of the views of Goethe, Oken, and the *Naturphilosophen*. He replaced the God of Christian tradition, who stood outside nature and was the creator of it, with Goethe's God: "There is nothing within, nothing without: but what is within is without."[17] Darwin's doctrine of descent provided a framework for a new *Naturphilosophie:* Monism; the unity of nature provided a scientific definition of God:

Every conception of God that separates Him from matter, and sets up in opposition to Him a sum of forces not of Divine origin, leads to amphitheism and polytheism. While Monism establishes the unity of the whole of nature, at the same time it demonstrates that only one God exists, and that this God reveals himself in a sum total of natural phenomena. While Monism bases all phenomena of organic and inorganic nature on the universal law of causality, and demonstrates that they are the result of efficient causes, at the same time it shows that God is the necessary cause of all things and the law itself.... In acknowledging all natural laws to be Divine, Monism rises to the highest and most exalted conception of which man, the most perfect of all animals, is capable: the conception of the unity of God and nature.[18]

There were three aspects of this Monistic world view: the general doctrine of development whereby "a vast, uniform, uninterrupted and eternal process of development obtains throughout all nature;"[19] the doctrine of biological evolution, which formed an essential part of general development; and the mechanism of natural selection, which gave an important nonteleological explanation for this development. A necessary corollary to Monism was a belief in abiogenesis:

If we do not accept the hypothesis of spontaneous generation, then at this one point of the history of development we must have recourse to the miracle of supernatural creation.... If we assume the hypothesis of spontaneous generation for the origin of the first organisms ... then we arrive at the establishment of an uninterrupted natural connection between the development of the earth and the organisms produced on it, and in this last remaining lurking place of obscurity, we can proclaim the unity of all nature, and the unity of her laws of development.[20]

Abiogenesis was the very cornerstone of his philosophy, for to "reject abiogenesis is to dismiss Monism from consideration."[21]

Haeckel also had strong empirical reasons for accepting the possibility of abiogenesis. By the 1860s, he had concluded that life resided at the lowest level in naked protoplasm without membranes or nucleus or any complex intrusions. In his *Generelle Morphologie* and the more popular *Natürliche Schöpfungsgeschichte,* and also in his "Monographie der Moneren," he claimed that the most primitive organisms or *Cytoden* were little but undifferentiated, anucleate, naked pieces of albuminous jelly, far simpler in construction than a simple cell. The result of such a view was, "the deep chasm which was formally and generally believed to exist between organic and inorganic bodies is almost or entirely removed, and the way is paved for the conception of spontaneous generation."[22] Genuine abiogenesis had thus become "a logical postulate of scientific natural history."[23] In 1868, these views received strong support from Huxley's discovery of *Bathybius haeckelii,* substantiating Haeckel's claim that the bottom of the ocean was covered in a universal *Urschleim.*

Haeckel distinguished clearly between heterogenesis and abiogenesis, and regarded the former as "only of subordinate interest for the history of creation." He claimed that continuous abiogenesis could never be disproved and that in the past, "under conditions quite different from those of today, the spontaneous generation which now is perhaps no longer possible, may have taken place."[24] In the context of a monistic world view such a statement makes little sense. If the most primitive organisms were as simple as Haeckel assumed, there was no valid reason to believe that spontaneous generation "is perhaps no longer possible." Such a statement could only be justified by assuming that primitive organisms were actually quite complex and the result of a long evolutionary history. The monistic world view prevented such an assumption, for it implied that all gradations of material from the simplest chemical atom to man exist in nature.

The German physicalists were in a dilemma. Having opposed spontaneous generation in the previous decade, partly as a reaction against the sterile metaphysical speculations of their predecessors, their new commitment to Darwinism led them directly into the problem posed by the origin of life. Faced by their own experimental evidence, which did not support spontaneous generation, they were also confronted by the uncomfortable fact that to deny its occurrence was both to resurrect vitalistic notions and, worse, to deny that all natural events were "subject to the laws of necessary causation." They were committed to the concept that "the science whose object it is to comprehend nature must proceed from the assumption that it is comprehensible, and in accordance with this assumption investigate and conclude until, perhaps, she is at length admonished by irrefragable facts that there are limits beyond which she cannot proceed."[25]

Before Darwin, the greatest dilemma facing them was to explain the apparent purposefulness and design in living organisms. Schwann, for example, was forced to account for this by reference to the original creator: "The ground of this adaptation does not lie in the powers, but in Him, who has so constituted matter with its powers, that in blindly obeying its laws it produces a whole suited to fulfil an intended purpose."[26] The fact that Darwin's mechanism of natural selection had accounted for such purpose without reference to the creator, led them to respond most favorably to this aspect of Darwinism.

Helmholtz remarked, in 1869, that the Darwinian theory "shows how adaptability of structure in organisms can result from a blind rule of a law of nature without any intervention of intelligence."[27] Some years later, the great popularizer of science, Emil du Bois-Reymond expressed this feeling in no uncertain terms:

To explain how material particles not directed towards any definite end should nevertheless cooperate to that end.... The possibility, however remote, of banishing out of nature this apparent adaptation to ends, and of everywhere setting up blind necessity in the place of final causes, is to be regarded as one of the greatest advances ever made in the world of thought.... That he has in some measure diminished that torture of the mind that tries to understand the universe, will be Charles Darwin's highest title to fame.[28]

Accepting Darwinism they thus faced the problem of abiogenesis. An assumption of an extraterrestrial origin of life was an obvious way out of the dilemma. By so doing they could deny the continuous occurrence of abiogenesis without, at the same time, being thereby forced to assume a divine creation. First seriously proposed by Hermann Richter in 1865,[29] the position was also adhered to by Frederick Lange, who felt justified in placing the problem of the origin of life within the transcendental domain:

We find ourselves in the process of *ad infinitum,* and this kind of "postponement" has at least the advantage that it brings the unsolved difficulty into good company. The origin of life thus becomes as explicable and as inexplicable as the origin of the world generally; it comes into the sphere of transcendental problems, and to transfer it into this sphere is by no means logically improper, as soon as natural science has good grounds within its sphere of knowledge to regard such a theory of transmission as relatively the most probable.[30]

Although Helmholtz himself was to support this notion in 1871, he did not consider it necessary in the "decade of simple protoplasm." If, indeed, the simplest forms of life were homogeneous lumps of simple protoplasm, which had, in the words of Haeckel, "arisen out of slime and are nothing but slime in different forms,"[31] then the origin of life is reduced to "an exceedingly difficult mechanical problem."[32] In other words, the physicalists would not be forced to face the dilemma of abiogenesis until the demise of the theory of simple protoplasm.

The British, with their strong geological tradition, took a different approach. They saw the question about the beginning of life in concrete geological terms and for this reason ignored it. This was not surprising for they had ignored the problem long before Darwin arrived on the scene.

Traditional accounts of the history of geology have assumed that before the work of Charles Lyell it was a speculative science and that the role of the geologist was to explain geological phenomena in supernatural terms. This pre-Lyellian geology, termed "castastrophism," was described as a belief that the earth's history was explicable by causes which were supernatural, nonactualistic—having no analogies with events of the present—and sudden or catastrophic. Recent work by Martin Rudwick has suggested the contrary; that many pre-Lyellian geologists believed these sudden catastrophic events "could be generated from an essentially gradualistic background, without any suspension of the ordinary laws of nature."[33] In the 1820s, for example, the French geologist, Élie de Beaumont, gave such a rational account of catastrophes. He explained that as the earth gradually cooled and shrank, stress areas built up in the earth's crust, which would suddenly buckle to produce mountain chains and other such phenomena.[34]

The problem of the first appearance of life in the fossil record, and the successive appearance of new species thereafter, presented immense difficulties for the geologist. It is simply a gross oversimplification to claim that the geologists accepted the direct creation of life, whether all at once in the beginning or successively over time. Naturally, some did believe this, but even so, it would be dangerous to conclude that their geology was subservient to their theological beliefs.[35] This claim seems to have arisen from Darwin's own insistence that there were only two means by which one could explain the appearance of life over geological time: by one or a series of special direct creations or by slow transmutation of species from the original organism or organisms.

Cuvier, the dominant figure in early "catastrophic" geology, did not believe that the catastrophes were universal or that it was necessary to recreate the species after every cataclysm. He believed that successive strata were populated by a migration inward from areas not affected by the catastrophe:

When I endeavour to prove that the rocky strata contain the bony remains of several genera, and the loose strata those of several species, all of which are not now existing animals on the face of our globe, I do not pretend that a new creation was required for calling our present races of animals into existence. I only urge that they did not anciently occupy the same places, and that they must have come from some other part of the globe.[36]

It was his pupil, Pierre Flourens, who first spoke of "a creation entirely destroyed and lost,"[37] and the British geologists who expanded Cuvier's doctrine into universal catastrophes. But even during this early period William Buckland, the famous British catastrophist, spoke of all major divisions of animal life being present at the beginning.[38] By the 1820s catastrophes were often seen as local events with thoroughly naturalistic explanations.

The empirical and naturalistic basis of this geology led them to ignore the problem of the cause of the beginning of life. Lyell, for example, was very concerned to distinguish between geology and cosmogony. The former was the science "which investigates the successive changes that have taken place in the organic and inorganic kingdoms of nature," and was not concerned with questions as to "the origin of things."[39]

There were two reasons for taking this approach. First, the earlier fossil remains were very complex, having nothing in common with Haeckel's cytods or the universal *Urschleim*. Their origin seemed inexplicable by natural means, especially that of abiogenesis. The problem was then attributed to the Divine act or simply ignored. In the 1860s, for example, both Darwin and Huxley took the former position. Huxley stated in 1860 that: "With respect to the origin of this primitive stock, or stocks, the doctrine of the origin of species is obviously not necessarily concerned. The transmutation hypothesis, for example, is perfectly consistent either with the conception of a special creation of the primitive germ, or with the supposition of its having arisen, as a modification of inorganic matter, by natural causes."[40] Darwin himself, in all editions of *The Origin of Species*, concluded by upholding the creation belief, speaking of "several powers having been breathed into a few forms or into one."[41] Privately, however, he believed no such thing, for in a letter to the botanist Joseph Hooker in 1863 he wrote, "I have long regretted that I truckled to public opinion, and used the Pentateuchal term of creation, by which I really meant appeared by some wholly unknown process."[42] The issue was obviously not an important one for him, for in all later editions of *The Origin* he never saw fit to change the term he had used for the first beginning of life.

The second reason sprang from the empirical traditions of mid-nineteenth-century British science. The popular concept of science, as expressed in periodicals, saw scientists following the inductive methods of Francis Bacon and that aspect of science expressed in Newton's dictum: *"Hypotheses non fingo."* Scientists were regarded as fact-collectors or, as *The Times* so succinctly expressed it, "we look to men of science rather for observation than for imagination."[43] It was clearly absurd to speculate on such matters as the beginning of life.

Many British scientists, such as Huxley and Darwin, patterned themselves after the teaching of the British positivist John Stuart Mill. The positivism of Mill not only allowed science to ignore first causes but also to accept supernatural events as the first cause. To British positivists causes were concrete events which preceded other events and which therefore were subject to experimental verification. Since the first cause was by definition not a phenomenon, then clearly it lay outside the domain of science. As Mill himself noted in his *System of Logic*: "When in the course of this enquiry I speak of the cause of any phenomenon, I do not mean a cause which is not a phenomenon: I make no research into the ultimate or ontological cause of anything."[44]

Mill was also opposed to Comte's assumption that belief in a designer reflected an early stage in man's thinking that would pass away as society progressed and came to apply the fruits of the scientific method to all facets of human existence. "Positive mode of thought is not necessarily a denial of the supernatural ... positivist philosophy maintains that within the existing order of the universe ... the direct determining cause of every phenomenon is not supernatural but natural. It is compatible with this to believe, that the universe was created, and even that it is continually governed, by an Intelligence."[45]

These views were reflected not only in Huxley's insistence that evolution and creation were compatible but also in Darwin's letters of the 1860s, when he wrote of the problem of spontaneous generation being "beyond the confines of science,"[46] and further that "it is mere rubbish, thinking at present of the origin of life; one might as well think of the origin of matter."[47] A Newtonian justification for such views is seen in Darwin's letter to Lyell of 1860. "He [Bronn] seems to think that till it can be shown how life arises it is no good showing how the forms of life arise. This seems to me about as logical ... as to say it was no use in Newton showing the law of attraction of gravity and the consequent movement of the planets, because he could not show what the attraction of gravity is."[48]

The origin of the individual species also presented problems to those pre-Darwinian geologists who were opposed to any supernatural explanations. There was simply no evidence to substantiate the views of Lamarck and Étienne Geoffroy Saint-Hilaire that the species had evolved from preexisting species. Most geologists were then content merely to state that they had come into existence. Such statements were not so much a conscious avoidance of the issue as a reflection of the type of explanation which then seemed applicable to geological science. Pre-Lyellian geology was dominated by the belief that a directional plan existed in the geological record. Rudwick claims that it would be far better to term this earlier geology "directional" rather than "castastrophic." This directionalism was brought about by a gradual cooling of the earth and an associated increasing complexity of organic life. The role of the geologist was to describe this progressive trend, *not* to dwell on sterile questions of mechanical causality. Geologists were thus content to describe the successive appearance of the fossil species and to ignore the problem of how they came into being. Clearly, however, the fossils' complexity ruled out any notion that they were formed by a process of spontaneous generation. In that sense, then, the geologists of the pre-Darwinian period were part of the general British hostility towards spontaneous generation.

One cannot deny that this British opposition was a result of the strong effect of natural theology on their biology and geology; but, after the 1820s, such natural theology was usually subservient to the geological evidence. Both these factors led the British to deny the possibility of ascribing the first appearance of life and of the successive species to a process of spontaneous generation. Nevertheless, by the 1860s, explicit denial of spontaneous generation gave way to silence over the issue. In 1869 this silence was to be broken, and the implications of the theory of evolution for the problem of spontaneous generation were brought into open debate.

Before this, however, the British were drawn indirectly into the debate over spontaneous generation by way of the continuing controversy over infectious diseases. In the nineteenth century, Britain and Europe were ravaged by four cholera pandemics, which reached England in 1831, 1848, 1853, and 1865 respectively. They accounted for approximately two hundred and sixty thousand deaths in England alone.[49] The major concern of public health officials in England was naturally directed toward the prevention of these periodic epidemics, which were seen to have spread inevitably from India across Europe into Britain.

Belief in the contagious nature of cholera and other diseases was widespread during the eighteenth and early part of the nineteenth century, but quarantine, the standard preventative measure for all diseases considered to be contagious, failed to check the onslaught of the early cholera pandemics and thus raised serious doubts as to the validity of the contagion doctrine. As we know today, it failed to prevent the entry of cholera into Britain because of the presence of nonsymptomatic carriers and because of the length of the incubation period, and also because many diseases are infectious before any symptoms appear. In addition, there was an exact parallel with parasitic worms in that contagionist doctrines could not explain the irregular susceptibility of persons to the disease or the parasite.

After the first cholera pandemic, anticontagionist doctrines came into vogue. Believing that the disease was a result of the deplorable social conditions of the poor, the anticontagonists attributed the disease to filth, overcrowding, sewage, lack of drainage, unwholesome food and water, and odors. As a result they were responsible for initiating attempts to legislate public health bills in order to improve the sanitary conditions of the English towns and cities. Being opposed to quarantine regulations, the anticontagionists received financial, moral, and political support from the rising class of merchants and industrialists who had suffered monetary losses when their goods had been quarantined, and who were opposed to government intervention in trade so long as they enjoyed the benefits of *laissez-faire* economics.[50] Since cholera, like typhoid and dysentary, is an infectious disease of intestinal origin, any attempts to improve sanitation would prove to be of inestimable worth in checking the spread of the disease. This is precisely what happened in Britain; public health officials led by John Simon, Edwin Chadwick, and others utilized improved methods of sanitation in their fight against cholera. By these means alone the death rate from cholera was reduced from seventy-two thousand in 1848, to twenty thousand in 1865. After the inception of the Public Health Act of 1875, the country escaped further cholera epidemics beyond local outbreaks in various ports.

John Snow's work of 1849 and 1855, when he showed that cholera victims had taken their drinking water from the Broad Street Pump in the City of London, and the work of William Farr in the epidemic of 1865, gradually led to the reemergence of contagionist theories. Very few of these contagionists believed the contagion to be a living organism, however. Joseph Lister, for example, who had introduced antiseptic techniques into the Glasgow Hospital in 1865—with dramatic results—remarked in 1871 that, "I do not ask you to believe that the septic particles are organisms," although he himself believed they were.[51]

Most British contagionists of the middle and late nineteenth century believed the contagia to be *nonorganismic* particles of one sort or another. All such beliefs implied that the disease could arise either by contact with the contagion or *de novo*. This belief, of course, *does not* imply any support for the doctrine of the spontaneous generation of *organisms*. Among those holding to such a view was Lionel Beale, a professor at King's College, who was not only opposed to heterogenesis, but, being a "vitalist," was necessarily opposed to abiogenesis. Beale believed that the contagion was of living material, but not a living organism. The contagia were "degraded forms of living matter, derived by direct descent from some form of human living germinal matter or bioplasm,"[52] which were disseminated in air, food, or water. Such particles existed in the blood and tissues of all organisms, and under normal conditions either divided to form other bioplasmata or differentiated into the component tissues of the organism. In cases of excess nutrients, however, the multiplication of the bioplasm became excessive, since the property of differentiation had no time in which to manifest itself. In this case the bioplasm could leave the blood stream and become the contagion. This excess production of bioplasm could also occur at the expense of the host tissue itself, and in that case bacteria might come to grow in it. The bacteria, it must be noted, were not produced by the host tissue. Clearly such a disease could occur either through contact with some external contagious bioplasm or *de novo* when suitable changes occurred in the host tissue. The specificity of disease was explained by the specificity of the bioplasm. It was postulated that, since specific particles of bioplasm differentiate into specific tissues under normal situations, they were also capable of initiating specific disease symptoms when conditions were not normal.

There is a close parallel here with the "theory of protoplasm" that dominated cell concepts in the 1860s. If life at its simplest level was manifested by naked clumps of homogeneous protoplasm, there was no reason to deny the ability of protoplasm (bioplasm) to act as disease contagion. More particularly, there was obviously a very close parallel between Beale's bioplasm and Darwin's theory of pangenesis. This theory was elaborated in his *Variation of Animals and Plants under Domestication,* which appeared in 1864. Here Darwin writes of cells dividing or throwing off "minute granules or atoms, which circulate freely throughout the system, and when supplied with proper nutrients multiply by self division, subsequently becoming developed into cells like those from which they were derived."[53] He even made reference to Beale's views when he likened his gemmules to "an atom of smallpox matter."[54] Similarly, Darwin's gemmules were specific, having an "elective affinity for that particular cell which precedes it," and, like Beale's bioplasm, their production was enhanced by changes in external conditions.[55]

Similar views were held by Dr. James Ross of Manchester. He believed that "the particles of contagious matter are, not withstanding their similarity to bacteria, not independent beings, but modified units of the individual from which they have become detached."[56] He drew a parallel between grafting and the contagion attaching itself to the host organism and quoted Charles Darwin to the effect that "when trees with variegated leaves are grafted or budded on a common stock, the latter sometimes produces buds bearing variegated leaves; *but this may perhaps be looked at as a cause of inoculated disease.*"[57] In an earlier book he mentioned that speculation on the origin of such contagious matter was "useless or mischievous," and pointed out that "we do not hesitate to adopt Dr. Beale's modification of Harvey's maxim: *Omne vivum e vivo.*"[58] Both Beale and Ross believed that disease contagia did arise *de novo*, but neither believed in any form of spontaneous generation. The use of the modified maxim is however amusing, since this was the very phrase used by the German Romantics in their support of spontaneous generation.

Chemical concepts of contagion, stemming from the earlier views of Liebig, were also popular at this time. Benjamin Richardson, for example, a leader in experimental pharmacology, saw contagia as "albuminous ferments," which excited chemical changes in the host body and by so modifying the host's chemistry caused a breakdown in tissue and the pathology and symptoms of the disease. New chemical contagia were produced as a necessary excretory product of this breakdown, and these products could excite similar occurrences if transmitted to other host organisms. Clearly such views were almost a direct copy of the views of Liebig, and it is not surprising that they still enjoyed wide popularity as late as the 1860s and 1870s. In 1870, Richardson remarked that:

The physical theory of the origin of the communicable diseases, is that the diseases are due to poisons which are organic, but are neither independently reproductive nor indestructible, nor derived from organisms distinct from the animals in which the diseases are developed. It accepts . . . that their action is purely physical on the body. It leaves undecided the hypothesis of fermentation or putrefaction.[59]

He later modified this theory of zymosis to a so-called glandular theory, which held that the contagion was a modified type of glandular secretion, which, when passed to a new organism, causes the production of "the same poisonous property as was possessed by the poisonous substance just introduced."[60] Again such a theory allowed for a *de novo* origin of the contagion, but as Richardson had earlier pointed out, "the formation of the poison although from organic substances, has nothing to do with the origin of life; the poison is a purely chemical change or organic substance."[61]

None of these views had any real bearing on the spontaneous generation debate, particularly since most of that minority who believed the contagia to be living organisms were not interested in the origin of such organisms. John Burdon-Sanderson, professor of physiology at University College, expressed this point of view when he wrote:

The inquiry must be approached from the pathological side. As naturalists, we may be interested in all the lessons to be learnt from bacteria; in studying the remarkable influence of the conditions under which they originate and grow ... or the question how they come into existence, as it were, out of nothing. But all this is beside the mark; at all events, beside our mark as pathologists. To us, who are concerned about disease, and have its prevention and cure as our ultimate object, the subject loses its interest the moment it becomes unconnected with bodily disorder.[62]

Yet some did discuss the problem of disease by reference to spontaneous generation. John Tyndall, who presents to the historian a confused panorama of beliefs: evolutionist, materialist, germ theorist, and opponent of heterogenesis, agreed with Lister that the contagious agent was a specific living organism. He seemed to hold to this view because of "the virtual triumph of the antiseptic system of surgery," which seemed to Tyndall to be based on "living contagia as the agents of putrefaction."[63] Lister, of course, was aware that an acceptance of living contagia did not necessarily imply the existence of a complete living organism. Tyndall was unique among the germ theorists, however, in concluding that the origin of the living contagion was important to the science of medicine, since, if the organism arose in the diseased body by heterogenesis, then medicine would have no ability to conquer such diseases: "But, judged of practicality, what, again, has the question of spontaneous generation to do with us? ... Now, it is in the highest degree important to know whether the parasites in question are spontaneously developed, or are wafted from without to those afflicted with the disease. The means of prevention, if not the cure, would be widely different in the two cases."[64]

This statement of Tyndall's seems to imply that a belief in the spontaneous origin of the disease contagion stands in direct opposition to a belief in the infectious nature of the contagion. In fact, of course, one could believe that bacterial diseases arose either by contact with a contagion or *de novo*, in exactly the same way that it was once believed that parasitic worms were derived either through contact with eggs or by heterogenesis from the host tissue.

On the other hand, Henry Charlton Bastian, professor of pathological anatomy at University College and Britain's foremost advocate of spontaneous generation, continually stressed that a belief in spontaneous generation was completely compatible with the majority view that disease was caused by nonorganismic contagia. Bastian recognized, as did everyone else, that there were two kinds of infectious disease: those, like smallpox and silkworm diseases, that were associated with parasitic organisms and those where this association was unproven. In both cases, however, he believed that the contagious agent, whether parasite or something else, was a direct product of chemical processes occurring within the host tissue. In the former case this did involve the heterogenetic origin of the disease organism. In both cases the disease itself was of a chemical nature, "owing to the corpuscles acting after the manner of mere dead ferments. They would incite certain changes in the fluids . . . and these changes would culminate in the re-evolution of new and independent organisms. . . . The organism incites the disease by setting up molecular changes, rather than by its direct multiplication as a living thing."[65]

Such chemical changes could or could not produce a living organism depending on the nature of the chemical change, and, in direct contrast to germ theorists, Bastian believed that the contagion was a result, not the cause, of disease:

Whether the ferment be an independent living organism, a fragment of not-living organic matter, or some mere physico-chemical influence, the initiative fermentative change is in each case a result of chemical action. And similarly with regard to "contagium," whether it be an independent living organism, an altered though living tissue element, a fragment of dead organic matter, a chemical compound, or even the most vague influence of a "set of conditions" which suffices to generate contagium *de novo,* we have in each case to do with gradually initiated chemical changes.[66]

Since the concept of nonliving contagia was the most accepted, it follows that the ideas pertaining to disease were not incompatible with a belief in spontaneous generation. In this very restricted sense one can claim that disease theories in mid-nineteenth-century England moved a little in the direction of a general sympathy with the doctrine of spontaneous generation.

In England the strongest opposition to spontaneous generation rested on the belief that protoplasm and the primitive organisms were not simple and were a great deal more complex than inorganic matter. Many opponents on this count were vitalists, but this was not necessarily the case. Herbert Spencer and the botanist William Thiselton-Dyer denied the possibility of a continuous spontaneous generation on the grounds of complexity but not from vitalistic beliefs.

One of the most influential voices in the debate was that of William Carpenter, professor of physiology at the Royal Institution. Carpenter's views reflect the impact of Helmholtz's "law of the conservation of force" on British physiology, but while Carpenter accepted Helmholtz's concepts and denied the existence of a unique and special vital force, he did not conclude from this that spontaneous generation was a possibility:

There can be no doubt that the present tendency of scientific investigation is to show a much more intimate relation than has been commonly supposed to exist between vital and physical agencies; and to prove that, while the former are of a nature altogether peculiar, they are yet dependent upon conditions supplied by the latter. And the more closely these phenomona are investigated, the more intimate and uniform does that dependence appear; so that we seem to have the general conclusion almost forced upon us, that the vital forces of various kinds bear the same relation to the several physical forces of the inorganic world, that they bear to each other; the great and essential modification or transformation being effected by their passage, so to speak, through the germ of the organic structure, somewhat after the same fashion that heat becomes electricity when passed through certain mixtures of metals.[67]

He did not conclude from this that the division between the organic and inorganic worlds was an arbitrary one, and that abiogenesis must then be expected on a priori grounds. The vital forces, although of a physicochemical nature and produced by the conversion of physical forces, nevertheless required the presence of a living organism to manifest the conversion. Such views precluded any possibility of spontaneous generation:

The pre-existence of a living organism, through which *alone* can heat be converted into vital force, is as necessary upon this theory, as it is upon any of those currently received amongst physiologists. And it is the specificity of the material substratum thus furnishing the medium or instrument of the metamorphosis, which . . . establishes, and must ever maintain, a well marked boundary-line between the physical and vital forces.[68]

Another vociferous opponent of spontaneous generation in the 1860s was Beale, whom Gerald Geison describes as the vitalists' "foremost and most vocal spokesman."[69] For Beale there was an absolute division between living and nonliving matter, or, as he expressed it, between living, active, formative matter on one side and lifeless, passive-formed matter on the other. In 1870 and 1871, with Huxley's and Haeckel's views so widely known and the belief in *Urschleim* so widespread, Beale produced two books violently attacking the mechanical concept of protoplasm and the mechanistic concept of life.

The fundamental question to Beale was: "Whether the arrangements of matter, the form of the living being, and the guidance of the forces in living things, are due to working of ordinary material forces, or to a power of an altogether different order.[70] Beale had no reservations about the answer to such a question. The idea of vital activity being explained by the conversion of physical forces seemed absurd: "Correlation is the abracadabraca of the long prophecized but still nonexistent science of mechanical biology."[71] "Life," he wrote, "is a power, force, or property of a special and peculiar kind, temporarily influencing matter and its ordinary forces, but entirely different from, and in no way correlated with, any of these."[72]

Supporters of vitalistic doctrines are necessarily opposed to any belief in abiogenesis, but not necessarily to heterogenesis. Such a position can be seen in the writings of John Hughes Bennett, professor of medicine at the University of Edinburgh. In 1862 he expressly denied the possibility of spontaneous generation, but at the same time it is clear that his views were compatible with heterogenesis, a doctrine which he later came to accept. He considered organic matter to be unique, composed of histogenetic molecules, precipitated out of organic fluids, and histolytic molecules produced by the breakdown of organic matter. Growth and development of organic tissues, attributed to the "successive formation of histogenetic and histolytic molecules" obeyed molecular forces acting in obedience to fixed laws.[73] Such laws, however, were not the simple laws of Newtonian physics, but unique to organic molecules.

By 1869 Bennett had come to accept heterogenesis. He saw a hierarchy of reproductive methods as one descended the animal scale: "The law of descent, therefore, from parents, *homogenesis,* which we recognize in the higher organisms, changes as we descend in the scale, first to *parthenogenesis,* where this direct descent is broken, and ultimately to *heterogenesis* in which it is lost."[74]

He did not regard the various living forms capable of being generated heterogenetically as distinct species, but rather forms which readily passed from one to another. He then went on to discuss at length the large number of observations which had failed to locate any or sufficient organic life in the atmosphere; the lack of proof that boiling will destroy life; the lack of proof that fermentation and purification depended on living germs rather than chemical phenomena as put forward by Liebig; and that such diseases as smallpox, scarlatina, measles, and typhus had an organic origin. Infusoria, it was concluded, "originate in oleo-albuminous molecules, which are formed in organic fluids, and which, floating to the surface, form the pellicle or proligerous matter. There ... the molecules, by their coalescence, produce the lower forms of vegetable and animal life."[75]

This vitalistic background to a belief in heterogenesis and a denial of abiogenesis was precisely that of Buffon in the eighteenth century. Indeed both Bennett and another British vitalist, Edward Parfitt, often mentioned that their views were similar to those of Buffon.[76]

Beyond Remak, Virchow, and Liebig, whose views have already been discussed, there were very few Germans at this time who were opposed to the concept of abiogenesis. This was the age of materialists and mechanists, all of whom supported the idea that abiogenesis was, at least, a theoretical possibility.

From what has been adduced so far, one would be hard pressed to conclude that, by the middle years of the nineteenth century, spontaneous generation was almost universally discredited. Indeed, one would have to conclude the opposite. The physicalists and materialists in Germany and, to a lesser extent, some of the British biologists had accepted a concept of a simple protoplasmic basis to life. In theory, the simpler the construction of primitive organisms, the more possible spontaneous generation becomes, whether by heterogenesis or abiogenesis. But, an acute dilemma had followed from these arguments over the nature of living matter. Abiogenesis had raised its head again and heterogenesis, being associated historically with vitalism, had retreated. As mentioned elsewhere this was an absurd situation.

A further aspect worthy of note concerns the experiments on spontaneous generation which were being conducted during this period. Few of those engaged in the debate over the nature of living matter carried out any such experiments. And, although the findings of Pasteur were widely quoted, the work of Schroeder, von Dusch, and others was hardly mentioned, except by Pasteur, who lived in a very different intellectual milieu. It was agreed, however, that no one had proved spontaneous generation to have occurred. The reactions to these experiments, which have been exaggerated out of all proportion by experimentally-minded modern scientists, is to be expected given the logical problem inherent in the controversy: they were of limited value in attempting to negate universally the concept of spontaneous generation.

In this account, several references have been made to the very different situation that existed in France. There, the doctrine of spontaneous generation was virtually destroyed during the first few years of the decade, and the position of the doctrine in France was a complete antithesis to its situation in Germany. To this contrast we now turn.

SIX

THE FRENCH *COUP*
DE GRÂCE

In the previous chapter we have pointed out that the doctrine
of spontaneous generation enjoyed strong support in the
seventh decade of the nineteenth century, particularly in Germany. In light of traditional accounts this is perhaps surprising, for
one normally associates this decade with the triumphant work of
Louis Pasteur, whose mastery of experimental techniques has long
been seen as responsible for the final overthrow of the vexing problem. Despite the work of Pasteur, however, the debate over spontaneous generation dragged on for another twenty years in Britain and
Germany, and only in France was the issue considered to have been
settled by him.

As demonstrated in an earlier chapter, support for spontaneous
generation in France was very short lived, so that the doctrine was
extremely unpopular even before Pasteur. To the French, Pasteur
simply provided the final experimental *coup de grâce* to a doctrine
which Georges Cuvier had virtually destroyed a generation earlier.
The doctrine of spontaneous generation had become associated with
the concepts of materialism, transformism, and the unity of type
through the work of Cabanis, Lamarck, Geoffroy Saint-Hilaire, and
Antoine Dugès. Cuvier had launched a vigorous campaign against the
views of Lamarck and Étienne Geoffroy Saint-Hilaire, culminating
with his famous confrontation with Saint-Hilaire in the late 1820s and
early 1830s. The ensuing victory of Cuvier rested to a large extent on
the scientific arguments he mustered on his own behalf, all of which
have been discussed. Briefly, they involved his views on harmony and
functional integrity of the organism, which led him to accept preformationist doctrines directly antithetical to any possibility of spontaneous generation. At the same time, these doctrines denied transformism by restricting variation within narrow bounds. Other features of
Cuvier's scientific attacks on Saint-Hilaire included his emphasis on
the discontinuities in the fossil record and his taxonomic scheme
which denied the unity of type in favor of four *embranchements*. Cuvier

92

was also seen to represent a more sober, cautious approach to science, then coming to vogue in France, by his emphasis on "positive fact" in opposition to the broadly philosophical ideals of science held by Saint-Hilaire and the German *Naturphilosophen*.

The controversy reappeared in France with the publication, in 1858, of a paper by Felix Pouchet, director of the Natural History Museum at Rouen, entitled, "Note sur des proto-organismes végétaux et animaux, nés spontanément dans l'air artificiel et dans le gaz oxygène." In this note Pouchet described the appearance of microorganisms in boiled hay infusions under mercury after exposure to artificially produced air or oxygen, and claimed thereby that germs in atmospheric air were not necessary to engender life in previously boiled liquids.[1]

This paper drew an immediate response from the influential French naturalist, Henri Milne-Edwards. Stressing the materialist basis of a belief in spontaneous generation, he argued that there was no proof that high temperatures destroyed the germs of infusions and that while chemists might synthesize the material of a living organism, the living organism itself could not be generated "without the assistance of a vital force."[2] Such an assumption, he argued, was supported by the geological record, which showed clearly that "since the creation until the present time, there has existed a non-interrupted chain of the possessors of this power which is passed on successively, and that brute matter cannot organize itself in such a way as to form an animal or plant."[3]

At the same time Jean Louis Quatrefages, Claude Bernard, and others attacked the experimental findings of Pouchet on the grounds that the heat resistance of germs was greater than previously believed, and that the atmosphere was filled with a large number of small organic spherical bodies.[4] Pouchet countered, in a series of papers, by denying that living organisms ever survived boiling temperatures, and by claiming that the particles in air were so "infinitely rare" as to render inexplicable the genesis of organisms "that one sees appearing in most cases in such prodigious numbers."[5] "In all our observations," Pouchet claimed, "we have never encountered either a single spore, a single egg of a *microzoaire*, or an encysted animalcule."[6] Such particles, Pouchet maintained, were "grains of wheat and granules of silica."

Of equal significance in this early exchange between Pouchet and so many of the "biological establishment," was the implication by Milne-Edwards that a belief in spontaneous generation implied an acceptance of materialism and denial of geological evidence. Pouchet not only denied that the geological record showed "a non-interrupted chain" of beings but disclaimed any pretence to "suppose that animals and plants will be produced solely by the action of general forces on which depend the chemical combinations in the organic world."[7]

This association of spontaneous generation with such extra-experimental issues was to play a very profound role in the final outcome of the debate in France. Earlier, the concepts of materialism and transformism had been linked with political, social, and theological issues at variance with the stability and well-being of the French state.

The years following the turmoil of the French Revolution saw a reaction against late-eighteenth-century atheism and revolutionary ideals. This reaction included a revival of Catholicism, which gained the political support of the Bourbon Monarch, who once again wished to forge a new society upon the union of throne and altar. The revival resulted in a polarization of French society between those opposed to any change in the status quo and those adhering to the ideals of revolution, liberalism, and democratic thought. Not only did Cuvier publicly associate his opponents with the atheistic and socially dangerous beliefs of the revolutionaries and with the materialism of the late-eighteenth-century *philosophes* and *idéologues*, but the father of French Romanticism himself, François-René Châteaubriand, had, in 1802, attacked the materialism of the revolutionary *idéologues*.[8] Since many of these figures of attack had publicly subscribed to the atheistic doctrine of spontaneous generation, the accusation of materialism directed toward Pouchet carried with it deep religious, political, and social overtones. To the French, the doctrine of spontaneous generation was part of a broad package of politically and religiously dangerous doctrines, and ought therefore to bear its share of guilt for the horrors of the recent past. In addition, belief in spontaneous generation was associated with the materialistic teachings of the newly-emerging and threatening German nation.

A generation later, at the time of the Pouchet-Pasteur debate, the situation was even more polarized than before. Not only were the forces of liberalism and democracy growing stronger all over Europe, but the revolutions of 1830 and 1848 had once more illustrated the threat posed by revolutionary ideals to the well-being of the French state.[9]

In the presidential election of 1848, Louis Napoleon sought for and received the support of the Catholics through his election manifesto, which promised that the state would protect the Catholic Church in France, grant freedom of education to the Church, and guarantee freedom and temporal power to the Pope in Italy. Since the Church controlled the vote of the newly enfranchised peasant voters,

Napoleon was returned by a landslide, and the later election of the Legislative Assembly resulted in a triumph for the Monarchists. The Monarchists and Catholics then formed a new party of the right, the party of the "Rue de Poitiers," with its antirepublican sentiments. The *coup d'état* of 1851 and the ensuing plebiscite resulted in the birth of the Second Empire, with Napoleon on the throne, supported by the Catholic Chuch. Most Catholics obviously agreed with Montalembert's statement: "A vote for Louis Napoleon does not signify approval of all he has done. The choice is between him and the complete ruin of France.... I believe that, in following this course of action, I am once more on the side of Catholics against the Revolution."[10]

In the complex political and religious events of the Second Empire, two aspects remained stable and of paramount importance. The first was the absolute necessity of Napoleon's retaining the support of the Catholic Church and the second was the regime's fear—and the fear of all factions of the Church—of republicanism and the forces of revolution. The Church and the State embraced in the face of the common enemy. The polarization was so complete that republicanism and opposition to either the Church or the State were regarded as synonymous. In this context, any attack on the Church or its doctrines was seen as an attack on the State, and vice versa.

Opposition to the Church, and thus the state, came not only from the political forces of republicanism, it came also from the growth of positivism, materialism, and atheism. All these were intimately linked to the scientific progress of the nineteenth century. For many, science became a sort of religion in its own right. Hippolyte Taine, for example, wrote: "The growth of science is infinite. We can look forward to the time when it will reign supreme over the whole of thought and over all man's actions."[11]

In the face of these attacks, the Church became more and more authoritarian and reactionary. In 1864 Pope Pius IX threw his weight behind the conservative and orthodox wing of the Church. The Papal Encyclical of 1864 condemned any suggestion that society could survive while practicing religious tolerance. To this encyclical was appended the "Syllabus of Errors", which included the notorious Error Number 80: "The Roman Pontiff can and must make his peace with progress, liberalism, and modern civilization and come to terms with them."

The dissension was further exacerbated by the appearance of Clémence Royer's notorious translation of Darwin's *Origin of Species*. A materialist, republican, and atheist, Madame Royer prefaced her translation with a lengthy diatribe against the Catholic Church, on which she blamed all of society's ills: "There is no conquest of the human spirit which has not infringed on their domain, not a discovery which has not beaten a hole in their system, which with great pains they have repaired each time, plastered and painted, filling the holes with paradoxes and supporting the creviced portions with sophisms."[12]

It is scarcely surprising that Darwinian evolution was regarded in France as a politico-theological doctrine allied with the forces that were attempting to overthrow Church and State.[13] Nor is it surprising that the majority of Darwin's French critics focused on the issue of spontaneous generation. For, besides its historical association with evolutionary theories, the French realized that Darwinism logically demanded an abiogenetic beginning of life. Since the essential beliefs of the Christian Church rested on Creation as well as Divine Providence, Original Sin, Incarnation, and Redemption, any support for spontaneous generation attacked the very core of Catholicism. Indeed, for the French, the presumed disproof of spontaneous generation by Pasteur led to their using Pasteur as the strongest argument against Darwin.

Two years after the Royer translation, Ernest Renan's sensational *Vie de Jésus* appeared. Rejecting all supernatural beliefs, he attempted to rewrite the life of Christ, based on historical criticism and scientifically justifiable events. He thus denied the resurrection of Christ:

The strangest rumours were spread in the Christian community. The cry, "He is risen," quickly spread amongst the disciples. Love caused it to find ready credence everywhere. What had taken place? For the historian, the life of Jesus finishes with his last sigh. But such was the impression he had left in the hearts of his disciples and a few devoted women, that during some weeks more it was as if he were living and consoling them.[14]

Against this awesome background the final French battle over spontaneous generation was played out. By associating Pouchet's experimental statements with many of these political, theological, and social issues, Milne-Edwards had once again made explicit the dangerous implications of the spontaneous generation doctrine. Not surprisingly, therefore, Pouchet was forced to debate these issues in the defense of his beliefs.

In 1859 Pouchet published his major work on spontaneous generation, *Hétérogenie ou traité de la génération spontanée.* It is not clear why Pouchet should have chosen to become embroiled in such a contentious issue. He was nearly sixty years old at the time and his major scientific publications had appeared in the 1840s. These were all related to problems of animal generation. They included *Théorie positive de la fécondation des mammifères* in 1842; *Théorie positive de l'ovulation spontanée et de la fécondation des mammifères et de l'espèce humaine* in 1844; and *Théorie positive de l'ovulation spontanée et de la fécondation* in 1847. Between 1847 and 1859 his only major work was a 650-page opus entitled, *Historie des sciences naturelles au moyen âge, ou Albert le Grand et son époque considérées comme point de départ de l'école expérimentale.* Neither is it clear at what point he became an advocate of spontaneous generation, for in these earlier biological works he seemed to deny its occurrence. In these works he advocated the same "fundamental laws of physiology," which included the conclusion that "in all the animal kingdom from the simplest organisms to man, generation occurs by means of eggs, and it is no less certain that they pre-exist at fecundation"; and that these eggs are produced by "a spontaneous and periodic ovulation."[15] Such eggs "were formed of microscopic vesicles more or less numerous, filled with a fluid in which were myriads of granules."[16] Clearly, such views reflected the then widely-held theory of exogeny put forward by Schwann, and indeed Pouchet drew attention to this similarity. Nevertheless, while numerous scientists at that time saw the parallel between the doctrines of exogeny and spontaneous generation, Pouchet remarked that only those organisms reproducing by fission or budding did so by means other than eggs. No mention was made of organisms arising spontaneously, and, indeed, in 1848 he endorsed Ehrenberg's claim that infusorians contained complex organs.[17] This is significant, since Pouchet's countryman, Felix Dujardin, at that time was arguing in favor of spontaneous generation on the basis of the simplicity of primitive organisms.

Surprisingly, perhaps, Pouchet's views in 1859 showed only a slight change from those expressed earlier. Now he used these same "fundamental laws of physiology" to justify a belief in spontaneous generation. But—and this is the most distinctive feature of his version of spontaneous generation—it was *not* the adult organisms which were thereby produced, but only their *eggs:* "Spontaneous generation does not produce an adult being; it proceeds in the same manner as sexual generation which, as we will show, is initially a completely spontaneous act by which the plastic force brings together in a special organ the primitive elements of the organism."[18]

This feature of his belief in spontaneous generation is one that has been overlooked in the traditional accounts of his debate with Pasteur, as it was also overlooked by his contemporaries. They also overlooked the vitalistic and providential character of his beliefs. Heterogenesis, he argued, was not the chance doctrine of the ancient atomists; instead, it implied that, "under the influence of forces still inexplicable, and which, as Cabanis said, will remain inexplicable, a plastic manifestation is produced, whether in the animals themselves, or elsewhere, which tends to group molecules and to impose on them a special mode of vitality from which finally a new being results."[19] This spontaneous generation, however, did not produce an adult but simply the initial stages of development, and "the same plastic force which takes place in the organization of animals and plants can also manifest itself in plant and animal debris."[20]

Pouchet's concern to show that his views on spontaneous generation had none of the atheistic and dangerous connotations usually attached to the doctrine came out clearly in his text of 1859, in which the first 137 pages were devoted to an historical and metaphysical justification for belief in spontaneous generation. In the second chapter, Pouchet came to grips with the religious arguments put forward against the doctrine of spontaneous generation. He agreed that the first origin of life was "a true spontaneous generation operating under divine inspiration," but to deny any further spontaneous generation was an "illegitimate fear, for if the phenomenon exists, it is that God has wished to use it in his design."[21] "Where is the verse in the sacred text," he asked, "which tells us that he imposed on himself never to resume his work? Or where is it said that after this rest, he has broken his moulds and annihilated his creative faculty."[22] He argued that God, having fashioned the germs of things, had also imposed laws of matter and of life which determined when the organizational forces gave rise to new beings. "The laws of heterogenesis," he insisted, "far from weakening the attributes of the Creator, can only augment the Divine Majesty."[23] In keeping with his vitalistic conception of spontaneous generation, Pouchet denied the abiogenetic production of life, for only living matter was endowed with the necessary *"force plastique."*

The succession of life on the surface of the globe links matter in a narrow circle from which it cannot escape. It is successively attracted and repelled by these incessant phenomena. But the organic particles, sometimes intimately united to form organisms, and sometimes free in space, are no less animated with a latent life, which appears only to wait for their grouping to be visibly manifested. It seems that for organic molecules, there is no death ... only a transition to a new life.[24]

In chapter six of *Hétérogenie*, Pouchet attempted to show that his views on spontaneous generation were compatible with current geological thinking. Denying that the geological record showed, in the words of Milne-Edwards, "a non-interrupted chain," he argued instead for a theory of successive creations. This aspect of the debate was important, since he thereby linked his theory with the orthodox geological theory and dissociated himself from Lamarckian or Darwinian transformism. Pouchet's arguments carried a greater cogency than one might expect.

He agreed with François Pictet that "the theory of successive creations is the only one which agrees with the law that the species are all different from one strata to another."[25] The origin of these successive creations was a mystery to the French geologists, and although some were content to ascribe their appearance to a direct creative act, most simply assumed they arose by natural and unknown means. The problem of the origin of these species was seen to be outside the scope of geology; geologists merely described their successive appearances and were not concerned with problems of causality. Pouchet's theory of heterogenesis provided such a mechanism, since the "plastic force" could be manifested in organic debris and give rise to the first primordia of the new creations. The geological opposition to the idea that complete fossil organisms could be produced by heterogenesis was thus countered by the argument that the egg, not the adult, was produced in this manner. He then claimed that it was irrational "to believe that this great work, so frequently repeated, ought to stop." Instead, he argued that, just as the intensity and universality of the geological catastrophes had diminished over time—the last universal catastrophe being that which built up the Andes and caused the Mosaic Flood—so the power of generating new beings "no longer attains the same proportions as in ancient times." Just as present "catastrophes" were limited to local minor upheavals, so the plastic force was now limited to the production of the eggs of infinitely small organisms.[26]

Pouchet also attempted to renew the claim that parasitic worms provided evidence for spontaneous generation. In spite of their "extraordinary migrations," Pouchet still argued "that it has not been demonstrated to everybody's satisfaction that some helminths do not arise spontaneously."[27]

I will be quite amenable to consider the cysticerci as tapeworm monsters, in which development is thwarted, and which, when accidentally carried into the gut, can acquire their normal adult form; but that is not to say that their initial origin was not by a spontaneous generation. Thus we make a large concession to the illustrious naturalist at Munich [von Siebold]. But from that to admit that all *Taenia* normally originate from cysticerci swallowed by mammals, there is an uncommensurable difference. Thus we change nothing in our thesis.[28]

Such was Pouchet's theory of spontaneous generation. What makes it particularly significant, given the sensitive backdrop to the debate, was its essentially vitalistic and providential nature and its association with geological catastrophism rather than with transformism. Pouchet emphatically opposed materialistic doctrines and continually denied the possibility of abiogenesis. It was manifestly absurd for his opponents to accuse him of holding evolutionary, materialistic, and atheistic ideas, but in the polarized political climate of the Second Empire that is exactly what transpired; Pouchet found his name associated with heresies which he had explicitly repudiated.

Pouchet also found his scientific methodology denigrated as "metaphysical," as proof of which his critics used the following unfortunate passage from the preface to his *Hétérogenie:* "When by meditation it was evident to me that spontaneous generation was one of the means employed by nature for the reproduction of living things, I applied myself to discover the method by which this takes place."[29] Referring to this passage, John Tyndall described Pouchet in 1878— somewhat hypocritically—in these terms:

Ardent, laborious, learned, full not only of scientific, but of metaphysical fervour, he threw his whole energy into the inquiry. Never did a subject require the exercise of the cold critical faculty more than this one—calm study in the unravelling of complex phenomena, care in the preparation of experiments, care in their execution, skilful variation of conditions, and incessant questioning of results until repetition had placed them beyond doubt or question. To a man of Pouchet's temperament the subject was full of danger— danger not lessened by the theoretical bias with which he approached it . . . it is needless to say that such a prepossession required a strong curb.[30]

In more recent times, William Bulloch, in his *History of Bacteriology,* also quoted this passage of Pouchet and added: "This was a dangerous start with a subject like spontaneous generation—so full of pitfalls."[31] As Gerald Geison and I have argued even more recently, such accusations are not substantiated by the evidence.[32]

As a result of these writings of Pouchet, the French *Académie des Sciences* established the Alhumbert Prize of 2500 francs, to be awarded "To him who, by well conducted experiments, throws new light on the question of so-called spontaneous generation." The prize was awarded to Louis Pasteur in 1862, on the basis of his justly famous essay, "Mémoire sur les corpuscules organisés qui existent dans l'atmosphère," which appeared in the *Annales des Sciences Naturelles* in 1861. Succeeding generations of scientists have agreed with the judgment of John Tyndall, who wrote of this essay that, "clearness, strength and caution with consummate experimental skill for their minister were rarely more strikingly displayed than in this imperishable essay."[33] From this justifiable praise, however, succeeding generations have also seen the outcome of the spontaneous generation debate in terms solely of these experiments. The modern historian of science, Thomas Hall, writes of Pasteur that he:

did not permit his religious convictions to influence his scientific conclusions. His canon as a scientist, he firmly stated, was to stand aloof from religion, philosophy, atheism, materialism, and spiritualism.... It was Pasteur's point that the material conditions of life had momentous implications for religion, but that these implications must not interfere with the objective interpretation of experimental results.[34]

In saying this, Hall is reiterating not only the commonly held view that the issue was solved by an objective appraisal of Pasteur's experiments, but also the myth that science is a self-contained, independent body of knowledge uninfluenced by extra-scientific events. Such a view posits scientists—or at least those of the stature of Pasteur—as persons who leave all preconceptions at the laboratory door and deal only with self-evident facts. Yet, clearly, both Pasteur himself and the outcome of the debate in France were greatly influenced by social, political, and theological forces beyond the laboratory door.

The initial confrontation between Pouchet and members of the French Academy occurred just as Louis Pasteur was becoming convinced that fermentation depended on the activity of living organisms. In a series of papers published in 1857 and 1858 he laid the basis of a belief in specific ferments by showing that the same medium gave rise to different fermentations, depending on the nature of the microorganisms sown into it; that brewer's yeast induced only alcoholic fermentation and that "lactic yeast" induced lactic fermentation. These conclusions stemmed from his earlier work in crystallography, which had convinced him that molecular asymmetry was intimately associated with life and could not be produced artificially by ordinary chemical means.[35] In a résumé of this early work, he wrote:

When I affirm that no artificial substance has yet shown molecular asymmetry, I refer to artificial substances made entirely of mineral elements or derived from substances not themselves asymmetric.... More than this: take any asymmetric substance whatsoever, and if you submit it to fairly energetic chemical reactions, you may confidently expect to see the asymmetry of the primitive group disappear.... Artificial substances do not have asymmetric molecules; I could not point out the existence of any more profound distinction between the products formed under the influence of life, and all others.[36]

By his own account, the question with which he began the study of fermentation concerned the source of asymmetry in its optically active products, in particular, amyl alcohol. This end product of lactic acid fermentation was traditionally assumed to arise directly from the optically active sugar, a view to which Pasteur could not subscribe. As he remarked in his paper of 1857: "The molecular constitution of sugars seems to me to be very different from that of amyl alcohol. If this alcohol, when active, originated from sugar, as all chemists agree, its optical activity would derive from that of the sugar. I am loathe to believe this."[37] Since Pasteur believed that the molecular structure of amyl alcohol differed too much from that of sugar for its optical activity to originate there, and since he had come to associate optical activity with life, it was natural that Pasteur should have concluded that fermentation depended on the activity of living organisms. In this first paper on lactic acid fermentation, Pasteur described the presence of a grey material, which increased in amount as fermentation progressed and which consisted, like brewer's yeast, of minute globules. He regarded this "lactic yeast," at least tentatively, as a living organism. In this paper he also wrote that the lactic yeast "originates spontaneously," but immediately emphasized in a footnote that the word was used, "to describe the fact, entirely leaving aside any judgement on the question of spontaneous generation."[38]

Pasteur's views on the physiological base of fermentation led him directly into the spontaneous generation controversy. Implicit in the notion of the specificity of fermentative microorganisms is a commitment to an ordinary sort of generation for them. Only if they arose by an ordinary sort of reproduction, it would seem, could they retain the specific hereditary properties which must account for the specificity of their actions during fermentation. On this ground alone, Pasteur may have been predisposed from the outset against the spontaneous generation of fermentative microorganisms, but the argument never emerges clearly and explicitly in his works. Apparently of more immediate concern to him was the allegation, stemming from Liebig, that yeast (and presumably other "ferments") was a by-product of fermentation rather than its cause. For those willing to accept yeast as living, but unwilling to assign it any causative role in fermentation, this suggested the possibility that yeast arose heterogenetically from the organic substances in the fermenting medium.

Writing in 1861, Pasteur traced his own interest in spontaneous generation to his work on fermentation, and more particularly to his recognition that the ferments were living organisms:

Then, I said to myself, one of two things must be true. The true ferments being living organisms, if they are produced by the contact of albuminous materials with oxygen alone, considered merely as oxygen, then they are spontaneously generated. But if these living ferments are not of spontaneous origin, then it is not just as oxygen that the gas intervenes in their production, but rather as a stimulant to a germ carried with it or already existing in the nitrogenous or fermentable materials. At this point, to which my study of fermentation brought me, I was thus obliged to form an opinion on the question of spontaneous generation. I thought I might find here a powerful support for my ideas on those fermentations which are properly called fermentations.[39]

In fact, to preserve and buttress his physiological theory of fermentation, Pasteur had to address the problem of the origin of the microorganisms he considered responsible for the process. For, if these organisms arose heterogenetically in a medium already undergoing fermentation, their causative role in the process could scarcely be maintained. But if they arose only from preexisting organisms or precursor "germs," their causative role could be more confidently affirmed.

He addressed the issue of spontaneous generation for the first time in February 1859, when he showed that a boiled fermentable medium, when isolated from the air or exposed only to calcinated air, did not ferment. Lactic yeast, he concluded, came "uniquely by way of the atmospheric air." "On this point," he remarked, "the question of spontaneous generation has made an advance."[40]

This paper witnessed the beginning of Pasteur's confrontation with Pouchet, for it prompted a letter from Pouchet, which has apparently not survived, though Pasteur's reply has. "The experiments I have made on this subject," he began, "are too few and, I am obliged to say, too inconsistent in results . . . for me to have an opinion worth communicating to you." Nevertheless, he repeated the conclusion he had just announced in his published note and advised Pouchet that, if he repeated his experiments with the proper precautions, he would see "that in your recent experiments you have unwittingly introduced [contaminated] common air, so that the conclusions to which you have come are not founded on facts of irreproachable exactitude." Thus, wrote Pasteur, "I think, sir, you are mistaken, not for believing in spontaneous generation, for it is difficult in such a question not to have a preconceived idea, but rather for affirming its existence."[41]

Beginning in February 1860, Pasteur presented a series of five notes to the *Académie* on the subject of spontaneous generation, which were brought together in his prize-winning essay, "Mémoire sur les corpuscules organisés qui existent dans l'atmosphère," published in the *Annales des Sciences Naturelles* in 1861. Recognizing that the existence of atmospheric germs was as yet unproven, Pasteur set out to show that air did contain living organisms and to deny that, "there exists in air a more or less mysterious principle, gas, fluid, ozone etc., having the property of arousing life in infusions."[42]

In the first series of experiments outlined by Pasteur in his

"Mémoire," he attempted to repudiate the contention of Pouchet that the "germs" in the atmosphere were only grains of wheat and silica. Pasteur examined the solid particles of the air, which he collected by aspirating atmospheric air through a tube plugged with gun cotton, and then dissolving the plug in a mixture of ether and alcohol. Although this treatment killed any organisms that might be present, the trapped particles had "clearly defined outlines much like the spores of common molds." Seeking better evidence, Pasteur then repeated the experiments of Theodor Schwann. He boiled a 10 percent solution of sugar in yeast water for two or three minutes and then filled the flask with calcinated air. After sealing the neck of the flask with a flame, the contents remained clear, "even after standing for 18 months at a temperature of 25° to 30°."[43]

Pasteur next carried out an experiment which clearly illustrated his experimental ability. He attached the flask of "sterile" sugared yeast water, prepared as outlined above, to a series of tubes so arranged that a small wad of gun-cotton charged with atmospheric dust could be made to slide into the hitherto sterile liquid in the flask.

A diagram of the apparatus is shown. The sealed neck of the flask was fitted into a glass tube, which enclosed a small tube (a), open at both ends, containing a wad of cotton. The glass tube was fitted into a T-shaped brass tube (R) provided with stop cocks, one stop cock being connected to the glass tube, one to a suction pump, and the third to a red-hot platinum tube. After evacuating and admitting "calcinated" air into the tube T several times, Pasteur broke the seal to the sugar-yeast flask and allowed the gun-cotton to slide in. He then resealed

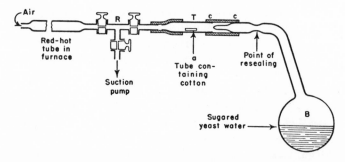

One of Louis Pasteur's early experiments on spontaneous generation, in which dust from the air is placed in contact with a fermentable liquid in an atmosphere of heated air. The diagram is from J. B. Conant, ed., *Harvard Case Histories in Experimental Science* (Cambridge, Mass.: Harvard University Press, 1957), vol. 2, in Case 7, "Pasteur's and Tyndall's Study of Spontaneous Generation," p. 509, and is used with the permission of the publisher.

the flask with a flame. In every case, growth appeared after 24 to 48 hours.

Finally, in order to counter the objection that the organisms may have arisen spontaneously from the organic matter in the gun-cotton, Pasteur substituted asbestos for the dust-laden gun-cotton and obtained the same results. With dust-free asbestos, on the other hand, the flasks remained clear. From these experiments, Pasteur concluded, "that all the organisms which appear in sugared albuminous solutions that have been boiled and then exposed to ordinary air have their origin in the solid particles suspended in the air."[44]

Pasteur then attempted to extend his conclusions to media other than yeast water, namely urine and milk. Pasteur claimed that boiled urine, deprived of atmospheric dust, could be stored indefinitely without alteration. Milk, however, heated in the same manner, invariably coagulated, this coagulation being associated with the appearance of microorganisms. Pasteur explained that this did not imply spontaneous generation, for, if either the duration of boiling increased or the temperature rose, the number of samples which showed coagulation decreased. Indeed, if the temperature of the milk was raised to 100°–112°C, no coagulation occurred. Pasteur concluded that the reason for this heat resistance was the alkaline nature of the milk.

Perhaps the most influential of all his experiments involved the famous swan-necked flasks. Pasteur drew out the necks of the series of flasks into very narrow extensions, curved in a variety of ways and exposed to the air. Without sealing these extensions he boiled numerous liquids in them, such as yeast water, urine, beet juice, and pepper water. When placed in calm air none of the contents of these open flasks fermented, while unboiled media in similar flasks did ferment. Pasteur concluded from these experiments, that the atmospheric dust had been captured in the sinous extensions and that the necks had thus prevented the contents of flasks from being contaminated.

By this time N. Joly, professor in the faculty of science at Toulouse, and his student, Charles Musset, had arrived at the same conclusion as Pouchet regarding the paucity of organic "germs" in the atmosphere.[45] They also argued that, if there were a universal dissemination of specific germs in the atmosphere, then the air must be so loaded with them as to appear foggy or even solid! In his essay Pasteur discredited this belief by showing that a limited quantity of air did not always produce microbial growth, that the atmosphere was not always filled with germs. This doctrine of " *la panspermie limitée*" received vivid support from the fact that it was possible to increase or decrease the proportion of flasks in which life appeared merely by exposing them to air in different localities. Fewer flasks showed microbial growth in the vaults of the Paris Observatory than in Pasteur's laboratory, and, most impressive of all, of twenty flasks opened at the foot of the Jura plateau eight showed microbial growth; of twenty exposed at 850 meters only five showed growth; and of twenty opened on the Montanvert glacier at an altitude of 2000 meters, only one flask underwent subsequent alteration.[46] He also criticized Pouchet's original experiment on spontaneous generation by showing that the mercury used by Pouchet could have been a source of contamination.[47]

This superb essay seriously damaged the doctrine of spontaneous generation as far as the origin of infusorians was concerned. As will be shown in the next chapter, later adherents to the doctrine were always accused of experimental error. Yet if the prize of the French Academy was awarded to Pasteur justifiably, it is equally clear that the adjudicators for this prize were far from an objective jury.

In the highly centralized bureaucracy of postrevolutionary French science, the outcome of any controversy was determined chiefly by the reaction of the Paris-based *Académie des Sciences*. This body typically responded to such controversies by appointing quasi-official commissions to adjudicate among conflicting doctrines, in order to arrive at presumably objective decisions, which thereby became the orthodox views of the whole French scientific community. To a very large extent, the victory of Pasteur over Pouchet was determined by the response of the two commissions set up in the 1860s to examine the question of spontaneous generation.

The first commission was set up to award the 1862 prize. This commission consisted initially of Geoffroy Saint-Hilaire, Antoine Serres, Milne-Edwards, Adolphe-Théodore Brongniart and Pierre Flourens, but before a judgement could be reached Saint-Hilaire died and Serres was dropped from the panel. Their places were taken by Claude Bernard and Jacques Coste. All of these members were declared opponents of spontaneous generation, producing a panel unanimously unsympathetic to spontaneous generation from the outset. Milne-Edwards and Bernard had already responded critically to Pouchet's paper of 1858; Brongniart and Flourens were disciples of Cuvier whose opposition to spontaneous generation was so well known; and Coste, a few years later, was to attack Pouchet's embryological concepts. Moreover, all were Catholics to whom the materialistic implications of spontaneous generation were anathema. According to Georges Pennetier, Pouchet and his collaborators who had entered the competition withdrew when some members of the commission announced their decision before they had even examined the entries, thereby Pasteur, received the prize uncontested, on the strength of his 1861 Mémoire.[48]

By 1864, however, Pouchet had forced the *Académie* to name a second commission on the basis of the results obtained in the Pyrenees. Here, in September 1863, Pouchet and his collaborators had exposed eight flasks of boiled hay-infusions to air at different altitudes and *all* had subsequently shown microbial growth. This, of course, was a direct contradiction of Pasteur's Alpine experiments and supported the claim of the heterogeneticists that organic material in infusions required only oxygen to form living organisms, particularly since no mercury was used in these experiments.[49]

By this time Pasteur had delivered another blow to the doctrine of spontaneous generation. Pasteur realized that all experiments so far had relied on organic matter which had been boiled, and thus were open to criticism in that boiling modified organic matter to such an extent that it was henceforth incapable of generating life. But, in March 1863, he finally succeeded in preserving inviolate two liquids—blood and urine—without heating them; he collected them directly from the veins or bladder of healthy animals and then exposed them only to germ-free air.[50] Pasteur always regarded this experiment as one of his most important and remarked in 1864: "No, there is no circumstance known today in which one can state that microscopic beings are produced without germs."[51] These results, he asserted, "carry a final blow to the doctrine of spontaneous generation."

The second commission of Milne-Edwards, Brongniart, J. B. Dumas, Antoine Balard, and Flourens was even more biased than the first! In addition, the social climate was even more hostile to spontaneous generation than before, since by this time the French had had time to contemplate the implications of Royer's translation of the *Origin of Species*. Also, by 1864, the infamous Papal Encyclical had appeared. In 1862, the historian and politician François Guizot had reminded the French of the implications of materialistic doctrines. "Under the blows with which they attacked Christian dogma," he remarked, "all the religious edifice collapses and all the social edifice shakes; the Empire, even the essence of religion vanishes."[52]

By 1864 the French were engaged in their attacks upon the Darwinian theory, most of them being directed against the question of the origin of life. Immediately after the appearance of the first edition of the *Origin of Species*, Ernest Faivre attacked the doctrine of spontaneous generation. "If we are logical," he stressed, "it leads us to two errors: the transformism of species in natural history, pantheism in philosophy." It was a problem, he added, which "excites at the moment the best minds, because it touches science, philosophy and religious beliefs."

While in France the doctrine of spontaneous generation is debated, in England and Germany one writes of the correlation of physical forces, the date of man's appearance on the earth, the origin, development and extinction of species; we demand calculations, observations, experiments, enlightenment for solutions to the grave problems of philosophy ... lifting us from the ... consideration of physical laws to that of general truths, which enlighten our reason and confirm our religious beliefs.[53]

In April 1864, *after* the appointment of the commission but two months *before* it met, Pasteur presented a public lecture to the social elite of Paris, at the Sorbonne, on spontaneous generation and its religio-philosophical implications. He opened his address with a list of the great problems currently agitating and dominating all minds:

Great problems are in question today, which keep all spirits in suspense: the unity or multiplicity of human races; the creation of man several thousand years or several thousand centuries ago; the fixity of species, or the slow and progressive transformation of one species into another; the eternity of matter ...; the idea of a useless God. These are some of the questions in which men indulge these days.[54]

In addition to these questions—indeed transcending them all—since it infringed directly or indirectly on each of the others and since it alone could be subject to experimental inquiry, was the question of spontaneous generation: "Can matter organize itself? Can organisms come into the world without parents, without ancestors? There's the question."[55]

After a brief historical sketch of the controversy—in which his aim was to show that the doctrine of spontaneous generation "has followed the developmental pattern of all false ideas"—Pasteur struck to the heart of the matter:

Scientific controversies are much more lively and passionate now because they have their counter-parts in public opinion, divided always between two great currents of ideas, as old as the world, which in our day are called materialism and spiritualism. What a victory for materialism if it could be affirmed that it rests on the established fact that matter organizes itself, takes on life itself; matter which has in it already all known forces! ... Ah! If we could add to it this other force which is called life ... what would be more natural than to deify such matter? Of what good would it then be to have recourse to the idea of a primordial creation, before which mystery it is necessary to bow? To what good then would be the idea of a creator God? ... Thus, admit the doctrine of spontaneous generation and the history of creation and the origin of the organic world is no more difficult than this. Take a drop of sea water containing some nitrogenous material, some mucus, some fertile jelly as it is called, and in the midst of it the first beings of creation take birth spontaneously. Little by little they transform themselves and climb from rank to rank, for example, to insects after 10,000 years and no doubt to monkeys and man after 100,000 years. Do you now understand the link which exists between the question of spontaneous generation and those great problems I listed at the outset?[56]

Clearly in such an environment a cold, objective, "scientific" assessment of Pouchet's claims became impossible. By publicly addressing himself to these issues, at this particular time, Pasteur made mockery of his often-repeated claims that "no religion, no philosophy, no atheism, no materialism, no spiritualism," belonged in science, and of his claim that the question of spontaneous generation "is a question of fact."[57]

If the climate was violently hostile to Pouchet, so were the commissioners. In November 1863, at the time when Joly and Musset asked for the appointment of the commission, Flourens expressed his views in no uncertain terms: "Pasteur's experiments are decisive. If spontaneous generation is real, what is required to obtain animalculae? Air and putrescible liquor. Pasteur puts air and putrescible liquor together and nothing happens. Therefore, spontaneous generation is not."[58]

In 1864 Flourens' attack on Darwinian transformism appeared. Entitled *Examen du livre de M. Darwin sur l'origine des espèces,* its major tenet was that Darwinism demanded the occurrence of spontaneous generation and thus could not be maintained since, "spontaneous generation is no more. M. Pasteur has not only illuminated the question, he has resolved it."[59] Although this text appeared before the commission met, Flourens was appointed to it. Similarly by this time the eighth volume of Milne-Edward's *Leçons sur la physiologie et l'anatomie comparée* had appeared. In it he presented a lengthy discussion of the controversy and attacked proponents of spontaneous generation with these words:

When the savage tribe of one of those isolated oceanic islands saw shipwrecked sailors for the first time, they thought that these strangers were descended from heaven or, like fishes, had arisen from the sea bed. They did not stop to think that they came from an unknown land situated beyond the horizon.... Partisans of spontaneous generation seem to me to reason in the same manner as these ignorant islanders.[60]

Not surprisingly Joly bitterly resented this type of attack.[61] Yet Milne-Edwards was appointed to the commission!

Other members of the second commission were no less hostile to the concept of spontaneous generation. Balard, one of the two members not on the first commission, had been Pasteur's mentor in chemistry and had long taken an active interest in his career. Indeed, according to Émile Duclaux, Balard actually suggested the "swan-necked" flask experiments to Pasteur.[62] The second new member, J. B. Dumas, was likewise a patron of Pasteur and may have felt a special bond with his protégé since he, like Pasteur, had been subjected to abuse at the hand of the German chemist, Justus von Liebig. Dumas, moreover, was an important political figure in the Second Empire, having been appointed minister of agriculture and senator by Napoleon, and later occupying the presidency of the Paris Municipal Council.

The commission eventually met with Pasteur and the heterogeneticists on 28 June 1864. The issue in question was the doctrine of *"panspermie limitée"*; whether, in the words of Pennetier, "one cubic decimeter of air taken in towns, on mountains, in the middle of the ocean or in the depths of caves . . . is always fecund."[63] The meeting broke up in disarray. Having been assured by Flourens that they would be free to carry out any experiments they so desired, the heterogeneticists discovered that, "M. M. Dumas, Balard, and Brongniart had imposed on us a program entirely different from that of M. Flourens, a program so restricted that it was confined *to a single* experiment, to an experiment of Pasteur which to us proves absolutely nothing. . . . We came to Paris not to glorify Pasteur's experiments but rather to demonstrate before you their complete uselessness."[64] Since the commission was only prepared to examine Pasteur's flasks of the *previously prepared* medium opened in different localities, Joly, Pouchet, and Musset apparently had no option but to "protest in the name of science" and withdraw.[65] Instead, they appealed to the public on two separate occasions. The first, on 28 June 1864, in the amphitheatre of the Paris Faculty of Medicine and the second, on 1 March 1865, in La Salle de la rue Cadet.

As in 1862, so in 1864, the commission observed only the experiments of Pasteur and once again supported his claims in a report which scarcely veiled its contempt for Pouchet and his colleagues. The commission's report was submitted to the Academy in February 1865, with the argument that its terms of reference had not been "to judge between two opposing doctrines," but rather, "to state the existence or not of facts which certain scientists affirm and others deny."[66] Contrary to all previous practice, Dumas and Milne-Edwards demanded that the Academy pass judgment on the report. Although some argued that such a procedure was unconstitutional, the chairman Joseph Decaisne called for a show of hands and pronounced the report adopted. Urbain Le Verrier protested that there was no such majority and, since Milne-Edwards wished to eliminate any ambiguities, a recount was taken. This time an absolute majority prevailed, although, according to the account of Émile Alglave, "tumult" ensued on account of numerous abstentions on procedural grounds!

This report, which in a very real sense saw the virtual end of the spontaneous generation controversy in France, was produced without any attempt having been made to refute the findings of Pouchet's experiments in the Pyrenees. Pasteur, during the commission, chose merely to repeat his own secure experiments with yeast, despite which the commission warmly praised his exactitude. Pasteur and members of the commission thus violated one of the fundamental precepts of the "experimental method" in not falsifying the experiments of an opponent. This violation is even more remarkable in the case of spontaneous generation since, logically speaking, opponents of the doctrine, being unable to prove that no organisms arise spontaneously, could do no more than falsify those experiments said to prove its occurrence. As Pasteur himself recognized on numerous occasions, "In the observational sciences, unlike mathematics, the absolute, rigorous demonstration of a negation is impossible."[67]

Pasteur usually chose to ignore Pouchet's findings in the Pyrenees, which were extremely significant since they were carried out in the absence of mercury, the presumed source of contamination in his 1858 experiments. Only once, in a note of November 1863, did Pasteur attempt to challenge the Pyrenees experiments. He criticized Pouchet and his collaborators for limiting the number of flasks to eight and on the really quite desperate grounds that they had broken their sealed flasks with a heated file, rather than with a pair of pincers as Pasteur had! "In this important detail," Pasteur added, "they have departed from my mode of operating."[68]

Had Pouchet been allowed to repeat his experiments with hay infusions, the commission members would have been in some difficulty. As would be realized in the next decade, Pouchet's medium contained heat-resistant spores, Pasteur's did not. Although this was not known in this decade, there is still no disputing the biased nature of the jury and no excuse for the uncritical acclaim its members accorded Pasteur's experiments.

Naturally, Pouchet and his supporters were not convinced. In 1871 Pasteur once again was forced to debate the issue as part of his defense of his theory of fermentation. The chemist Edmond Fremy and the botanist Auguste Trécul both considered yeast and bacteria to be engendered from "albuminous matter." Although Fremy, in 1864, had rejected "without hesitation the idea of spontaneous generation," by which he meant, "the creation of organized beings, even the most simple, from elements not possessing a vital force,"[69] he claimed that "*corps hémiorganisés*" (albumen, fibrin, casein, etc.) may give rise to organized beings: "By reason of the vital force they possess, they undergo successive decompositions, giving birth to new derivatives, and engendering some ferments. Such productions are not due to a spontaneous generation but to a pre-existing vital force in hemiorganized bodies."[70]

In December 1871, after an address by Pasteur to the Academy attacking Liebig's views on fermentation, Fremy took the floor to criticize Pasteur's views. Admitting that he himself had had similar views in 1841, when, as he added somewhat patronizingly, "Pasteur had hardly entered science," he went on to claim that "under different circumstances, a single albuminous substance can form different ferments." He further asserted that, "in the production of wine, it is the juice of the grape itself, which, in contact with air, gives birth to grains of yeast, by the transformation of albuminous matter."[71] He questioned also whether the atmosphere contained sufficient germs to engender fermentation whenever grapes were exposed to air. Pasteur immediately responded by pointing out that germs exist on the surface of grapes as well as in liquids and air.[72]

Eight days later Trécul addressed the Academy on the same subject. Believing lactic bacteria "to be formed by modification of albuminous matter," he argued that the following changes occur during fermentation: "Albuminous matter changes into bacteria or directly into alcoholic yeast, or into *Mycoderma*."[73] Such transformations, he added, "only take place during the active stages of fermentations and never before fermentation begins."[74] Pasteur expressed surprise "to see the question of spontaneous generation being brought up again," and asked whether Fremy would admit to error, "if I can demonstrate that the natural juice of grapes, exposed to air devoid of germs, cannot ferment or give rise to organized yeast?"[75]

In October 1872 he carried out such an experiment. Into flasks of boiled and filtered unfermented grape juice, he poured one of three fluids: water in which grapes had been washed, water in which grapes had been washed and which was subsequently boiled, or juices extracted from uninjured grapes. Only the unboiled washing water solutions fermented, clearly illustrating to Pasteur that "the yeast which causes grapes to ferment in the vintage tub comes from the exterior and not the interior of grapes."[76] This brilliant experiment, similar to his earlier demonstration that natural urine could be preserved without heating, further convinced the vast majority of French scientists that spontaneous generation was a chimera. The debate dragged on, however. Fremy and Trécul continued to insist that yeast arose either from albuminous matter or by transformation from other types of organisms. Pasteur continued to answer their criticisms, much to the chagrin of Balard and Duclaux, who considered the episode closed and to have been a complete waste of valuable time— time that would better have been spent studying diseases of beer with which Pasteur was then engaged.

Études sur la bière, which appeared in 1876, had little or no impact on the spontaneous generation debate. The French needed no further convincing! However, perhaps partly in response to Fremy's and Trécul's criticisms, Pasteur now de-emphasized the role of atmospheric germs in fermentation and wrote: "Fermentation cannot take place in the juice of crushed grapes if the must has not come into contact and been mixed with particles of dust on the surface of the grapes, or of the woody part of the bunch."[77]

On 10 February 1878, the illustrious Claude Bernard died. While sorting out Bernard's unpublished papers, his pupil Arsène d'Arsonval, located a notebook containing some experiments on fermentation. The notebook was passed to the chemist Marcelin Bertholet, who then had it published in the *Revue des Cours Scientifiques* for 20 July 1878. "The formation of alcohol," the paper concluded, "is independent of all cells." "The ferment does not originate from exterior germs," but rather, "the alcohol is formed by a soluble lifeless ferment in ripe or rotting fruits."[78] Pasteur was horrified. Not only had Bernard attacked his biological theory of fermentation, but, more to the point, Pasteur saw in it a defense of the doctrine of spontaneous generation:

Although the expression spontaneous generation of yeast does not occur in any part of the Bernard manuscript, it is implied very explicitly many times. In his philosophical and physiological thinking, Bernard very often allowed his thoughts to wander at random more than one thinks and more than he would admit himself. Kind and aimable by nature, living in this elite world of the French Academy where spiritualistic ideas dominate, he often forced himself, both in conversations and particularly in his writings, to be extremely cautious, an attitude in accordance with the rigour of his scientific method. Only those scientists with bold spirits would adhere to a philosophy which they would be powerless to establish themselves. Therefore, I am not surprised to find in the [unpublished] Bernard manuscript a theory of spontaneous generation and the conclusion that grape ferment does not originate from external germs.[79]

Such an assumption was not justified. But Pasteur was so convinced of the necessity of the presence of living organisms to produce fermentation that any suggestion of a chemical causal agent appeared to imply a spontaneous generation of the organisms. He thus set out to show, once again, that yeast cells are necessary for fermentation. Taking immature grapevines on which yeast had yet to form, he wrapped them in sterile cotton. When ripe the grapes on these vines did not ferment unless exposed to the open air. "The question of a soluble ferment is then judged," he concluded. "This ferment does not exist where Bernard believed it to be."[80] The year was 1878, by which time German and British scientists had discovered the existence of heat-resistant bacterial spores, had begun work on cell division, and were rapidly concluding that contagious diseases were initiated by specific living organisms. As will be shown, all of these events led to the final disavowal of spontaneous generation in Britain and Germany. Despite the continuing criticisms of Fremy and Trécul, however, most Frenchmen were convinced that spontaneous generation had finally been discredited by experimental evidence.

It is my contention, however, that despite the wizardry of Pasteur's experimental skills, his victory was achieved because Frenchmen were already convinced of the impossibility of spontaneous generation. His experiments did not persuade those supporting the doctrine to change their minds. Although it is ridiculously exaggerated, Trécul made a valid judgement when he wrote:

Pasteur judges the quality of an experiment only from the results obtained. Every experiment is valid when it opposes heterogenesis.... Pasteur believes that if 999 experiments were favourable to spontaneous generation, and only one was not, only this last experiment would be considered valid. But one success in a hundred is enough for us.[81]

Pasteur's impact on the spontaneous generation debate incorporated more than his experimental attacks on the issue. Through the historical introduction to his 1861 essay, Pasteur has convinced generations of scientists and historians of science that the issue had always revolved around the questions with which he was concerned; that the sequence of events leading up to his own work had stemmed mainly from the experimental work of Redi, Needham, Buffon, Spallanzani, Schwann, Schroeder, and von Dusch. This strangely incomplete historical account may well have been an intentional strategy on Pasteur's part, for it seems to accord with his less-than-thorough attack on the issue of spontaneous generation.

Gerald Geison and I have pointed out that, while he was prosecuting his case against Pouchet, and although his attack on spontaneous generation was compatible with his own religious, social, and political views, he nevertheless believed in the possibility of abiogenesis and attempted to produce life artificially in the laboratory.[82] Such a belief stemmed from his conception of an "asymmetric force," the intervention of which he considered essential to the production of asymmetric molecules and, hence, of life. As early as 1852 he had tried to bring asymmetric forces to bear upon crystallization by means of magnets and he had suggested that, to produce "the immediate, essential principles of life," it was necessary "to manufacture some dissymmetric forces, to resort to the actions of a solenoid, of magnetism, of the asymmetrical movements of light, to the actions of substances themselves dissymmetric."[83]

The line of demarcation of which we speak is not a chemical question. . . . it is a question of forces. Life is dominated by these asymmetrical actions of whose developing and cosmic existence we have some inkling. I urge even that all living species primordially in their structure, in their external forms, are functions of the cosmic asymmetry. Life is the germ and the germ is life. Now who may say what might be the destiny of germs if one could replace the immediate principles of these germs (albumin, cellulose, etc.) by their inverse asymmetric principles. The solution would constitute in part the discovery of spontaneous generation, if such be in our power.[84]

But these remarks were not made until 1883, after the fall of the Second Empire. Before the fall and during the spontaneous generation debate, Pasteur had remained virtually silent over this issue. What needs to be stressed, however, is that Pasteur came into the debate in the face of a curious scientific dilemma. On the one hand, his work on fermentation led him to discount the possibility of heterogenesis; on the other, his work on crystallization and asymmetry led him not only to believe in the possibility of abiogenesis but also to attempt to produce such a conversion in the laboratory.

It was probably this scientific dilemma which led Pasteur into a less-than-thorough attack on the issue of spontaneous generation. This was manifested in numerous ways, including his violation of the experimental method during the debate with Pouchet, which Geison and I have outlined. Other instances of Pasteur's seemingly unscientific behavior may also be attributed to this dilemma.

In the first place, Pasteur remained outside the debate over Darwinism and evolution. This was curious, since his work was being used both to attack Darwin and to support the religious, political, and social ideals which Pasteur espoused. One would surely have expected him to have used such arguments in his debate with Pouchet, particularly when Pouchet was claiming that it was his own views, and not those of Pasteur, that were compatible with religious and scientific orthodoxy. Indeed, as the debate progressed, Pouchet put more and more emphasis on the argument that a belief in heterogenesis was not compatible with evolutionary atheism. In 1862, he wrote the geologist Jules

Desnoyers of his beliefs that "geologists are necessarily and irresistibly heterogeneticists," and that "to prove that a given century is not and will not be able to produce new organisms, is tacitly to advance that in the past all is derived from a single and unique creation."[85] The contents of this letter were merely a forceful restatement of the views put forward in his text of 1859. Pouchet claimed, also, in this letter, that the solution to the problem of spontaneous generation could only come from an examination of creation in its entirety and not from chemists examining the contents of their phials. In 1864, following the appearance of Roger's translation of *The Origin of Species,* he affirmed again that "the fixity of species, as put forward by Flourens, is the most important and demonstrated fact in natural history, notwithstanding the assertions of Darwin."[86] In 1868, in a preface to Pennetier's *L'origine de la vie,* he again claimed that: "Heterogenesis is a logical consequence of the appearance and ascension of organized beings at the earth's surface.... Since, it is evident that a new fauna and flora has arisen in each great period of upheaval, it becomes evident also, that life has many origins, and that in each of these periods it is necessary that these were numerous spontaneous generations."[87]

Pasteur chose virtually to ignore this issue of evolution and the origin of life. He mentioned Darwin only once by name. This occurred in his *Études sur la bière* of 1876, in which he discussed the belief that some microorganisms can transform themselves into yeast and give rise to fermentation. Such a view, he argued, "had lost more than gained ground abroad, in spite of the growing favour of Darwin's system."[88] This was the only reference to Darwin that he ever made.

Aside from his political address to the Sorbonne in 1864, Pasteur also remained virtually silent over the question of the origin of life. In the papers presented to the *Académie* in February, May, September, and November 1860, which culminated in his prize-winning essay of the following year, he mentioned the problem of the origin of life only once. At all other times he insisted that: "It is not a question of religion, of philosophy, of any system whatsoever ... it is a question of fact. And you will notice, that I have not the pretension to assert that spontaneous generation never exists. In subjects of this type one cannot prove the negative.[89]

But Pasteur did not, in fact, hold to these statements. In his dealings with Pouchet he conveniently ignored his statement: "In subjects of this type one cannot prove the negative." He *did* assert that spontaneous generation never exists, but he was careful to limit his discussions to the experimental facts springing from his work on fermentation.

Limiting his historical account of spontaneous generation to these experimental issues had the effect of bolstering the impact of his own experimental work. Amazingly enough, all traditional accounts of the spontaneous generation issue have been based on this account. One can mention William Bulloch's *History of Bacteriology,* and J. B. Conant's *Harvard Case Histories in Experimental Science* as two examples, but almost all biological textbooks find room for some aspect of Pasteur's story. It is clear that this historical introduction, when tied into Pasteur's experimental work, led succeeding generations to see the issue only in these terms. They, like Pasteur, reached back into the eighteenth and early nineteenth century to find the precursors to Pasteur's experimental work on infusions. They therefore told the now-familiar story of Redi, Needham, Buffon, and Spallanzani, in the eighteenth century, and Schwann, von Dusch, Schroeder, and others in the nineteenth.

In this historical introduction, Pasteur managed to ignore most of the key issues in the controversy, many of which were extremely detrimental to the concept of spontaneous generation. He ignored, for example, the experimental work in the early 1850s on the origin of parasitic worms, which had triumphantly concluded that spontaneous generation was not a viable explanation of their origin. Indeed, it was only a few years prior to Pasteur's involvement in the spontaneous generation issue that the *Académie* had set a prize essay on the problem of the development and transmission of parasitic worms. Pasteur also ignored the demise of the blastema theory of free cell-formation rooted in the work of Schwann. The theory of free cell-formation—the origin of cells out of a structureless cytoblastema—was used by some scientists in the 1840s to justify a belief in spontaneous generation, and it is clear that a reaction against spontaneous generation was a factor in Robert Remak's and Rudolph Virchow's views on cell continuity, first put forward in the 1850s. Although the impact of the famous Virchow dictum was minimal during these years, partly because there was no absolute demonstration of its validity and partly because the cell theory was changing to incorporate a theory of simple protoplasm, one might have expected Pasteur to have at least mentioned their writings. Pasteur also failed to mention the association of spontaneous generation with German *Naturphilosophie* and Lamarckian transformism, both of which were decidedly out of favor in mid-nineteenth-century France.

By behaving in this way, and by remaining virtually silent in public over his work on dissymmetrical forces, Pasteur managed, consciously or unconsciously, to steer the argument away from those scientific issues which were incompatible with his own theological, social, and political ideology.

By limiting his public work to those experiments which he had

reason to believe would discredit the doctrine of spontaneous generation, and by limiting his historical introduction to this work, he managed to direct a frontal attack on the issue of spontaneous generation
without appearing to directly contravene his own pronouncements
that "science ought not to worry about the philosophical consequences
of its work."[90] In other words, Pasteur did not worry about the
philosophical consequences of his work, simply because he took care
to limit the public work to those areas in which the consequences were
in tune with his own philosophy.

It is curious to see that Pasteur's attitude changed gradually after
the collapse of the Second Empire and with the rise of the secular and
more liberal Third Republic (1871–1919). Indeed, in the very year of
Louis Napoleon's abdication, Pasteur returned to these experiments
on asymmetry and life. Even before his address in 1883 to the Chemical Society of Paris, where he first admitted that he had attempted to
produce "the immediate, essential principles of life" in the laboratory
thirty years previously,[91] he had begun to qualify his earlier statements that spontaneous generation did not exist, and increasingly
admitted the difficulty of disproving it experimentally. In 1874, he
stated that the gap between the organic and mineral worlds could be
bridged by the action of asymmetric forces, an achievement which,
"would give admittance to a new world of substances and reactions
and probably also of organic transformations." "It is there," he continued, "that it is necessary to place the problem not only of the
transformation of species but also of the creation of new ones."[92] Four
years later in a short paper entitled "Sur l'origine de la vie," he made
the cryptic statement that, "I have looked for it [spontaneous
generation] for 20 years without discovering it. No, I do not judge it
impossible."[93]

The changed political and theological climate in France is also
apparent from a report of this address which appeared in 1885. After
reviewing Pasteur's work, the author merely commented:

The propositions maintained by M. Pasteur in this conference will certainly
surprise those who admire the scientific rigour always evident in his observations and experiments. They show that alongside the experimenter whose
methods are so severe, there is in M. Pasteur a bold and passionate
philosopher, who, once outside his laboratory, is not afraid to soar beyond the
facts and to launch out into the region of fertile hypotheses.[94]

Whatever the motives and subtle moves of Pasteur, one fact remains clear. Pasteur virtually convinced the French in the 1860s that
spontaneous generation was a dead issue, and as a result the main
focus of discussion in the 1870s was centered on Britain and Germany.

There, the concept of simple protoplasm and the Darwinian theory of evolution provided the impetus for a continuing belief in spontaneous generation. Curiously enough, however, by the end of that decade, when the British and Germans had been convinced that spontaneous generation was an impossibility and thus all speculations on the origin of life were futile, the French began to discuss it for the first time. The rise of republicanism after 1870 may well have been the crucial factor which allowed the French to finally discuss the contentious issue without reference to a creator.

In 1879, Felix Isnard attacked philosophies based upon religion, in his belief that one must "submit to rigorous proof of reasoning and to accept as truth only that which is demonstrated by science."[95] Reason, he argued, forces us to accept abiogenesis and "admits equally the possibility of heterogenesis, that is to say the spontaneous origin of many and new inferior organisms from organic and living bodies or from dead substances formally organized and more readily organizable than mineral matter."[96] In traditional French style, he then remarked that "transformism is a consequence of the spontaneous origin of life."[97]

In the same year, Edmund Perrier, one of the early neolamarckians, also admitted that "spontaneous generation is the foundation of the doctrine of evolution."[98] Three years later a dispassionate review of cosmozoan theories appeared in the *Revues Scientifiques*, in which it was admitted that "no deductive reasoning can prove that it [heterogenesis] is *a priori* impossible."[99] The positivist, E. Littre, while remarking that "positivist philosophy has no need today, anymore than in the past, to adhere to spontaneous generation,"[100] did so from his position that experimental evidence is "the only thing that I know."[101] In 1888 the palaeontologist Albert Gaudry, delighted in Darwin's *Origin of Species*, an experience, as he put it, similar "to sipping a small glass of delicious liquor."[102]

But theological opposition to spontaneous generation and evolution was still widespread. The priest, H. de Valroger, saw spontaneous generation as a necessary aspect of atheism and stated that: "The action of the creator has been necessary for the production of the first living beings."[103] But he also remarked, somewhat mysteriously, that Christian theology was no longer interested in spontaneous generation. In 1878 a book of essays by Paul Émile Chauffard, professor of medicine at Paris, appeared. It included such essays as "De la finalité dans les êtres-vivants et de la doctrine de l'évolution" and "La science et l'ordre social," in which spontaneous generation and Darwinian transformism were attacked. Linking both doctrines with revolutionism, atheism, materialism, and modernism, he saw the biological sciences "leading to general doctrines whose subversive application will lead to the ruin of a civilization imbued with spiritualism."[104] "What becomes of a country without God," he cried, "what more beautiful and profound cry than that of our fathers: God and the King!" French science, he urged, must reassert its old values and recover its spiritual traditions.[105]

Religious attacks on spontaneous generation were common in Britain and, to a very limited extent, in Germany. But only in France were proponents of spontaneous generation subjected to such vitriolic rhetoric as: "Those who defend it are sons of the 18th century who strayed into the 19th. Deluded and blinded; ignoramuses whom nothing can convince; *provocateurs* of seditious crises and even anarchists."[106]

The unique theological, social, and political climate of the French Second Empire provided the background against which the French support for Pasteur's work must be judged. Already opposed to spontaneous generation on theological and political grounds, French scientists saw the brilliant work of Pasteur as clinching the issue. Only after the fall of the Empire and the rise of the Third Republic did the French begin to discuss the issue in terms similar to those used by the British and Germans. But by the end of the 1870s, when a more liberal climate allowed such discussions to take place, the controversy was coming to an end even in Britain and Germany.

SEVEN

THE LAW OF GENETIC CONTINUITY

THE EIGHTH DECADE IN BRITAIN AND GERMANY

I n the 1870s belief in spontaneous generation, now restricted to questions over the beginning of life and the origin of microorganisms, was finally discredited by the discovery of heat-resistant bacterial spores, by the growing popularity of the parasitic "germ" theory of disease, and by the work of the German cytologists. The latter group showed that life was a continuous phenomenon, brought about by the mechanism of nuclear continuity from cell to cell and from generation to generation. It was these events which finally established "the law of genetic continuity" as a basis for all further research in the biological sciences, a law which had no place for the unpredictability inherent in the doctrine of spontaneous generation.

The victory of those opposed to spontaneous generation was not apparent in the early years of the decade. This was especially the case in Britain, where poor microscopic techniques and lack of interest in the life histories of microorganisms continued to provide problems for those opposed to spontaneous generation. The widespread belief that boiling killed microbes led to the assumption that any experimental "proof" of spontaneous generation was due to experimental error. Then, the role of those opposed to the doctrine was to show what these errors were. It was precisely the difficulty of showing such errors that enabled Henry Charlton Bastian, professor of pathological anatomy at University College, London, to mount an impressive defense of spontaneous generation in the 1870s. This defense was based on two premises: that boiling destroyed all life and that there was no valid proof of the existence of invisible germs in the atmosphere. In his first major paper of 1870, in which he noted the appearance of various microbes "in previously homogeneous solutions containing none of them," he argued:

But although microscopical investigation enables us to adduce evidence of just the same kind in elucidation of the mode of origin of certain low organisms, as we possess in explanation of the mode of origin of crystals, this evidence is not deemed adequate in the case of organisms. A living thing has been supposed to be a something altogether different, incapable of arising out of a mere collocation of matter and of motion; and, therefore, under the influence of this theoretical assumption, whilst chemists and physicists have thought that they could in a measure account for the genesis of crystals by reference to the affinities and atomic polarities of the ultimate constituents of such crystals, they have, for the most part, declined to adopt a similar mode of reasoning in order to account for the appearance of the minutest living specks in solutions containing organic matter.[1]

Yet, as Bastian pointed out, "so far as actual scientific evidence goes, [there is] almost as good reason for a belief in the universal distribution of invisible 'germs' of crystals, as there is for our belief in the universal distribution of invisible 'germs' of monads and bacteria."[2] Both concepts were equally hypothetical; no more was known about one than the other.

Despite such arguments and the experiments presented in this early paper deemed to illustrate a spontaneous generation of numerous microbes; despite his criticism of his opponents' reasoning that, "no matter what the temperature to which the solutions and the hermetically sealed flasks have been exposed . . . if Living organisms are subsequently found in the solutions, then they or their 'germs' *must* have been able to resist the destructive influence of such a temperature, simply because Living things have been found, and because it is *assumed* that they cannot be evolved *de novo*"[3]—despite all this, the British remained indifferent to Bastian's claims. Before 1873, when Bastian's experiments were substantiated by those of the famous physiologist Burdon-Sanderson, the British were convinced that Bastian was guilty of experimental error. Henry Moseley, for example, remarked in 1872 that, "it seems very unlikely that Dr. Bastian's results will be confirmed."[4]

In these early years, however, Bastian did manage successfully to embroil the British in the spontaneous generation controversy by associating it with debates over contagious disease and the origin of life. As pointed out in chapter five, the establishment of disease contagia did not dispense with the problem of spontaneous generation simply because very few physicians accepted that the contagia were living organisms or their "germs." Bastian was opposed to the parasitic theory of disease, and in a lengthy paper of 1872 argued that "the contagious element, whether living or not living, operates by its power of initiating similar changes which in their turn may or may not cause the evolution of new organisms."[5]

The refusal of the British to be drawn into the debate over the beginning of life, which had been raging in France and Germany during the preceding decade, came to an abrupt end in 1868. The American philosopher and Unitarian clergyman, Francis Ellingwood Abbot, in a spirited criticism of Herbert Spencer's *Principles of Biology*, argued that any biological treatise which still involved a creator, was "incompetent to yield an adequate philosophical basis for a general science of life."

The development theory must stand or fall with the theory of spontaneous generation. Logic permits no other conclusion. . . . It is very far from "immaterial" to the integrity of the development theory whether we believe that life first appeared *with* or *without* special miraculous creation. . . . The development theory is philosophically worthless, if it cannot altogether dispense with the help of that kind of agency, the assumption of which is its chief objection to the antagonist theory. It is bound to fill up the chasm between the organic and inorganic. . . . Law without miracle is the faith of science.[6]

He argued forcibly that scientists could not ignore forever the problem of the origin of life, and that once the issue was raised, "the burden of proof is transferred to the advocates of universal homogenesis, who must explain the apparition of the first organism, which *ex hypothesi* had no parents, as best they can."[7] It was this review which provoked Spencer to write his 1870 article on spontaneous generation, which was included subsequently as an appendix to his *Principles of Biology*.

Abbot's viewpoint was taken up by Gilbert Child, lecturer in botany at St. George's Hospital, London. In 1864 he had written a paper on the production of primitive organisms and had concluded that, although the experimental evidence dealing with the question of spontaneous generation was inadequate, nevertheless, "if the balance inclines to either side it is rather in favour of the heterogenists."[8] By 1869 he had concluded that the question was not resolvable by experimental means and that "we must have recourse to certain general considerations" in order to come to any conclusion on the issue. Such general considerations included the theory of evolution: that "heterogeny is part and parcel of the process of evolution," and that it was illogical, "to believe in the latter and deny the former."[9] "There must have been a stage in the development of the universe when the earliest forms of organic life were evolved from some special collocation of inorganic elements by the continued operation of the laws already in action."[10] In 1872 Bastian added his voice in support of Abbot and Child and became the central figure in the ensuing controversy. As a pathologist he was led to accept the possibility of spontaneous generation by knowledge in three areas that seemed to provide support for such a view: the cell theory, contagious disease, and parasitic worms.

In 1870 he wrote in defense of the "protoplasmic theory" in opposition to the growing emphasis on the inherent complexity of the cell. "The power of displaying vital manifestations," he wrote, "has, in fact, been transferred from definitely formed morphological units to utterly indefinite and formless masses of protoplasm."[11] In addition, he was a materialist and on many occasions expressed the view that life was simply emergent, the result of the organization of ordinary matter. Organisms "have been arbitrarily named living bodies," he wrote in 1872, whereas, "living things are peculiar aggregates of ordinary matter and of ordinary force which in their separate states do not possess the aggregate of qualities known as life."[12] Not surprisingly, given his views on protoplasm and the cell theory, he denied Virchow's dictum and subscribed to the concept of endogenous cell formation.[13]

Bastian still considered that parasitic worms provided support for the doctrine of spontaneous generation. Twenty-one years after Küchenmeister had initiated a decade of experimentation in parasitology, Bastian admitted that their origin could be explained without reference to spontaneous generation, but he also maintained that "the truth of such facts does not veto the possibility of the occasional independent heterogenetic origin of some of these parasites."[14]

Bastian argued in his books, *The Beginnings of Life* and *Evolution and the Origin of Life,* published in 1872 and 1874 respectively, that evolutionists must accept the abiogenetic origin of life. He maintained that it was not only illogical for evolutionists to accept a creation of life, but also that the continual presence of primitive forms demanded a continuous replenishment by spontaneous generation. This view, that primitive beings evolve into higher forms and thus that the earth would soon become depleted of such organisms without a continuous replenishment, was of course a restatement of the Lamarckian viewpoint.

Bastian also held views similar to those of Ernst Haeckel. The most primitive organisms or "Ephemeromorphs" were considered to be simple, transitory, and variable forms on which natural selection could not act. It was only after these forms had become so complex that such transitory and spontaneous variations became impossible during the life of each individual, that homogenetic production occurred and natural selection came into play. Once the continuous occurrence of spontaneous generation is admitted, then, "both the existence and protean variability of the lowest organisms are at once readily explained."[15]

In his 1874 text, Bastian elaborated more thoroughly on two aspects of his work that had been criticized. First, that his experiments that were said to prove heterogenesis had no bearing on the problem of abiogenesis, since to prove that organisms might be generated from a rich organic soup was not to prove also that they were capable of generation from simple inorganic substances. Second, that although abiogenesis might well be a necessity to the theory of evolution, it need not occur continuously.

The first criticism, that heterogenesis had no bearing on the problem of abiogenesis, was refuted by Bastian in a most strange fashion. He argued that, since there was only an arbitrary difference between the organic and the inorganic, thus there was no essential difference between heterogenesis and abiogenesis. He then assumed that any conclusions drawn from experiments on heterogenesis could equally well apply to situations involving abiogenesis—in other words, that the experimental proof of heterogenesis necessarily involved the proof of abiogenesis!

The process by which new living protoplasm comes into existence amongst this dead organic material, would be, for them [evolutionists] as much an instance of its new independent origin as if the process had occurred in the midst of mere inorganic elements. The term Archebiosis is therefore applicable to the process that must take place in our ordinary flask experiments where we deal with dead organic matter, just as it is also applicable to those more primordial combinations which first gave birth to living protoplasm on the surface of the earth.[16]

In answer to the second criticism, Bastian argued that to deny the present occurrence of abiogenesis and to assume that it had occurred only in the past was to "promulgate a notion which seemed to involve an arbitrary infringement of the uniformity of nature." "Evolution implies continuity and uniformity . . . it seeks to assure us that properties and tendencies now manifest in our surrounding world of things, are in all respects similar to those which existed in the past."[17] On the surface this was a very clever argument, for, in Britain, Huxley had clearly linked the biological theory of evolution with geological actualism. It appears that most British evolutionists simply refused to commit themselves. In a sense, they were justified in remaining undecided over the possibility of a continuous heterogenesis, particularly after the respected Burdon-Sanderson had lent support to those experimental findings of Bastian reputed to prove spontaneous generation. But Sanderson's experiments occurred after the British had begun to lose interest in the origin of life question. The failure of the British evolutionists to assume that abiogenesis was only an event of

the past may have resulted from the widespread belief that protoplasm and the most primitive beings were extremely simple. The
British debate over the origin of life took place at the very time when
Bathybius haeckelii was oozing its way through the universal *Urschleim* in
the depths of the ocean.

For the next few years the British discussed the problem of the
origin of life. In France and Germany, where the debate had been
crucial in the preceding decade, the debate was one in which the
materialists and the mechanists were ranged against those with more
vitalistic views. In Britain, however, this was only partially true since
most British naturalists were not prepared to admit that the evolutionary theory did necessitate a continuous occurrence of spontaneous generation. Indeed, of the British evolutionists, only Alfred Wallace and Thomas Stebbing accepted the argument of Child and Bastian that evolution demanded an acceptance of a continuous spontaneous generation.

Alfred Wallace greeted Bastian's views with enthusiasm. He was
at that time contemplating the bombshell that Lord Kelvin had introduced into the Darwinian debate. William Thomson, Lord Kelvin,
had attacked the idea that the earth's history spanned an enormous
length of time, concluding that the sun had been in no position to
sustain life until the last 20 to 25 million years, a time span that
seemed insufficient to account for the evolution of living things,
expecially considering the slow process of natural selection. Indeed,
the situation was made even more difficult by Darwin's insistence that
the variations upon which evolution was based must be numerous,
minute, and continuous, since any single large variation would quickly
disappear by being blended with more normal forms.

Wallace saw in Bastian's concept of the continual spontaneous
generation of "Ephemeromorphs" a much needed mechanism to increase the rate of evolutionary change, thereby "relieving the theory
of natural selection from some of its greatest difficulties and neutraliz-
ing some of the most serious objections that have been brought
against it." The evolution of Ephemeromorphs was seen to illustrate
"laws of progressive development," which resulted in a rapid and
early progress of life and under whose influence:

An immense variety of low forms of animals and vegetables would soon
people it [the earth]. It is fair inference too, that if such complex organisms as
Ciliated Infusoria, Rotifers, Nematoids and even simple Acari, can be developed independently of the slowly modifying influence of natural selection,
the same laws of development will continue to act a subordinate part much
higher in the scale, and, by assisting natural selection in its work, may have
enabled a much more rapid progress to be made.[18]

The Reverend Thomas Stebbing, evolutionist, naturalist, and pioneer in amphipod systematics, saw spontaneous generation "filling the gap in the order and continuity of nature, which is so puzzling without it." In the same vein as Bastian he argued that the simplest organisms continually advance as natural selection and variation take place, so that a continual replenishment of new and simple organisms would be necessary.[19]

Wallace was opposed to materialistic doctrines, believing that, although "protoplasm may not contain a peculiar form . . . it does manifest some other properties than the ordinary physical properties of its elements."[20] He therefore accepted the concept of a continuous heterogenesis, not a continuous abiogenesis, a stance contrary to those of Child, Bastian, and Stebbing.

Many of the other evolutionists, such as Darwin and Huxley, were content to extricate themselves from the dilemma by remaining essentially uncommitted over the question of spontaneous generation. Prior to 1870, Huxley did not view the origin of life as an important question and shared with Darwin the view that a belief in evolution was consistent with a belief in a creation of the first forms of life. In 1870, however, with the discovery of *Bathybius* and the simple nature of protoplasm rendering abiogenesis less improbable than before, he admitted that a belief in abiogenesis was an "act of philosophical faith."

I must carefully guard myself against the supposition that I intend to suggest that no such thing as Abiogenesis ever has taken place in the past or ever will take place in the future. . . . All I feel justified in affirming is, that I see no reason for believing that the feat has been performed yet . . . and if it were given me to look beyond the abyss of geologically recorded time to the still more remote period when the earth was passing through physical and chemical conditions which it can no more see again than a man can recall his infancy, I should expect to be a witness of the evolution of living protoplasm from not-living matter.[21]

In this paper, Huxley looked to pathology for a possible example of heterogenesis. He saw in corns, warts, and cancers, structures which were only morphologically distinguishable from parasitic worms, "the life of which is neither more nor less closely bound up with that of the infested organism." From this comparison he postulated that there might be "a kind of diseased structure, the histological elements of which were capable of maintaining a separate and independent existence out of the body," and in which "the shadowy boundary between morbid growth and Xenogenesis [heterogenesis] would be effaced."[22]

A few British "vitalists" became embroiled in the issue of the origin of life and, like their French and German counterparts, were led to reject evolution. The physiologist William Carpenter was one. As described in an earlier chapter, he had not only denied the possibility of abiogenesis but even seemed to favor the special creation of races of organisms and thereby, apparently, to deny evolution. "Thus we are obliged to go backwards in idea from one generation to another; and when at last brought to a stand by the origin of the race, we are obliged to rest in the Divine Will as the source of those wonderful properties, by which the first germ developed the first organism of that race from materials previously unorganized."[23]

J. H. Stirling, a Scottish doctor and student of Hegel, was drawn into the debate in opposition to the materialistic notions of T. H. Huxley. It is not surprising to find him stating a belief in immaterial forces in opposition to Darwin and the physicalists' doctrines, "neither Molecularists nor Darwinists then, are able to level out the difference between organic and inorganic."[24] More importantly, perhaps, L. S. Beale also mentioned the evolutionary problem in his descriptions of materialism and the nature of life. Beale regarded the activities and origin of his "bioplasts" to be incomprehensible on the basis of physico-chemical laws. "Will anyone believe," he asked, "that a fact so constant is meaningless, a result of accident to be explained by subtle influences, or natural selection?"[25] In one of his notoriously biting and sarcastic comments he added:

If we accept the statements that have been made, it may be regarded as certain, that at no very distant period an artificially evolved living being will triumphantly proclaim its own independent origin, and from its own experience relate to us a story of its evolution, traced back through various stages to its immediate construction out of the non-living, when the lifeless molecules of the inorganic so arranged themselves as to develop from their lower forms the higher vital mode. Then will the sceptic regret his unbelief, and joyfully become a convert to the New Philosophy.[26]

J. B. Pettigrew, a supporter of specific organismic disease etiology, also mentioned in passing that, since spontaneous generation was a phase of the evolutionary progress and since variation was only possible within narrow bounds, the theory of evolution was invalid.[27]

In Britain, unlike the situation in Europe, there was no polarization between the materialistic evolutionists on the one hand and the vitalist opponents of evolution on the other. This climate enabled people like Spencer and William Thiselton-Dyer to extricate themselves from the dilemma by arguments which seem to us far more rational than others put forward at the time. Both of them opposed abiogenesis on the grounds that the simplest forms of life were not of a simple construction and argued that such organisms were themselves, like all others, a product of a long evolutionary progress.

Thiselton-Dyer, professor of botany at the Royal College of Science, in Dublin, argued that, to accept an abiogenetic origin of life, "involves as much difficulty as the supposition of an absolute limit between the organic and inorganic worlds." The theory of evolution implied to him that life arose from nonlife by a series of "insensible gradations, the continuity of what is living with what is lifeless, of what is called vital with what is called physical."[28]

The interval which the evolutionist is modestly content to conceive deductively bridged, is as nothing to the leaping powers of the so-called heterogenist who boldly widens the gap and passes easily from ammonium tartrate to a Penicillium.... A believer in spontaneous generation is not indeed really an evolutionist, but is only a vitalist minus the supernatural; the special creation which the one assumes is replaced by the fortuitous concourse of atoms of the other.[29]

Similar views were held by Herbert Spencer and Henry Moseley, who in later years became professor of human and comparative anatomy at Oxford and served as a naturalist on the *Challenger* expedition. Spencer held that organisms were built up of "physiological units," consisting of large protein substances linked together. These units were "more minute, more indefinite and more inconsistent in their characters, than the lowest rhizopods." Such units were held to evolve from inorganic molecules by the process of natural selection, but the actual production of a living organism was seen as an enormously long process. There was no absolute commencement of life: "The affirmation of universal evolution is in itself the negation of an absolute commencement of anything."

I do not believe in the "spontaneous generation" commonly alleged ... and so little have I associated in thought this alleged spontaneous generation which I disbelieve, with the generation by evolution which I do believe, that the repudiation of the one never occurred to me as liable to be taken for repudiation of the other. That creatures having quite specific characters are evolved in the course of a few hours, without antecedents calculated to determine their specific forms, is to me incredible.... The very concept of spontaneity is wholly incongruous with the conception of evolution.[30]

As will be shown, arguments very similar to these were finally responsible for the demise of the spontaneous generation issue in the middle of the twentieth century.

Notwithstanding the views of Spencer and the British vitalists, some influential British evolutionists were ready to accept the possibility of abiogenesis. Such a stand seemed further justified when, in January 1873, the famous and respected British physiologist, J. Burdon-Sanderson, reported that he had watched Bastian perform experiments with hay and turnip-cheese infusions and found that he agreed with Bastian that "infusions can be prepared which are not deprived, by an ebullition of from 5 to 10 minutes, of the faculty of undergoing those chemical changes which are characterized by the presence of swarms of bacteria."[31] Sanderson reported that neutral hay infusions, neutral and acid turnip infusions, and neutral and acid turnip-cheese infusions swarmed with bacteria three days after having been boiled at 100°C. It was significant that bacteria appeared in the acid as well as in the neutral media, for Pasteur had suggested earlier that alkaline and neutral media might, as he put it, "resist sterilization." As far as acid media were concerned, however, it was generally agreed that boiling destroyed the life in them.

Even before these rather crucial experiments of Bastian and Sanderson, however, George Bentham, in an address to the Linnaean Society, noted that spontaneous generation "is still supported by so many naturalists whose opinions are entitled to consideration . . . that it is likely to be long maintained as a subject of controversy, without any further much more definite result."[32]

The widespread belief that boiling at 100°C for a few minutes was sufficient to destroy life necessitated opponents of spontaneous generation concluding that any appearance of microbes in a previously boiled medium was a result of experimental error. But now, since Sanderson, who had no a priori belief in spontaneous generation, had agreed with the findings of Bastian, it became necessary, for the first time, to take Bastian's experimental claims more seriously; they could no longer simply be ignored.

William Roberts was among the critics of the Bastian-Sanderson experiments. He suggested two sources of error in them: first, that an accidental reflux of air during sealing of the boiled flasks could introduce germs into the flasks and, second, that not all the contents of the flasks were exposed to the boiling temperature. To correct these errors Roberts suggested that a cotton plug be inserted into the flasks to prevent reflux and that the flasks should be heated in a water bath rather than over an open flame. In this way frothing and the adherence of particles to the vessel walls would be prevented.[33]

In June 1873 Burdon-Sanderson repeated the earlier experiments with Roberts' modifications and obtained the same positive results. He then introduced further modifications: lengthening the boiling time at 100°C and subjecting the media to high pressure, thereby raising the boiling temperature to a maximum of 102.68°C. This change of technique was necessitated by the unproven suggestions of Pasteur, E. Ray Lankester, and others that a temperature of 100°C might not kill all bacterial life. Sanderson obtained inconsistent results with this modified experiment, but, nevertheless, he felt able to claim that, "The chance that such a liquid will breed bacteria is diminished either by slightly increasing the temperature to which it is heated, or by increasing the duration of the heating. Thus it appears to me quite probable that if a sufficiently large number of flasks were heated even to 102°C, some of them would still be found to be pregnant."[34] Despite these remarks, Sanderson refused to be drawn into a general statement in support of spontaneous generation, to the understandable annoyance of Bastian. Indeed, one month later Sanderson remarked: "The particular experiment in question is not an instance of it [spontaneous generation], and that no argument founded on it in favour of the doctrine is of the slightest value."[35]

The debate over the Bastian-Sanderson experiments focused around two issues: either germs had been introduced into the flask by experimental error, or boiling at 100°C was insufficient to kill all bacteria. The Sanderson experiments of June 1873 added credence to the second possibility, and it was this question which next occupied the attention of Bastian. In March 1873, for example, he had drawn attention to experiments reported in his text of 1871, *The Modes of Origin of the Lowest Organisms*. In these experiments bacteria were inoculated into a boiled saline medium known to sustain but not engender bacteria. In flasks heated to above 140°F no bacteria appeared, but they did appear in flasks heated to a lower temperature. "These experiments seem to show," wrote Bastian, "that even if bacteria do multiply by means of invisible gemmules, as well as by the known process of fission, such invisible particles possess no higher power of resisting the destructive influence of heat than the parent bacteria themselves possess."[36] As far as Bastian was concerned the appearance of life in a medium that had been boiled at 100°C was ample proof of the existence of spontaneous generation, if temperatures of 100°C would kill all life. That was the crucial question and the concern about the effects of temperatures higher than 100°C seemed immaterial to Bastian. He believed that longer and higher boiling temperatures merely delayed the subsequent appearance of bacteria and accounted for the barren solutions which seemed to result from such modified treatment.

Unfortunately, however, the debate was further thrown into confusion by the findings reported by Ray Lankester and Joseph Lister. The former denied that bacteria had appeared in the turnip-cheese infusions after their having been boiled at 100°C for 10 minutes.[37]

Lister, whose antagonism to spontaneous generation is attributable to his success with antiseptic surgery, completely misunderstood the logic of the situation when he claimed that, when only one of many flasks of unboiled aseptically obtained milk remained free of bacteria: "Such success was as clear evidence against the hypothesis of spontaneous generation of organisms as if all the glasses had remained free from them." The assumption that a temperature of 100°C does kill life, led him to conclude that "all instances of so-called spontaneous generation had been due simply to imperfect experimentation."[38]

By 1873 the first part of Ferdinand Cohn's *Untersuchungen über Bacterien* had appeared. It created little interest in Britain since the studies simply supported the idea that a temperature of 100°C kills bacteria and that any subsequent appearance of bacteria was due to experimental error. Cohn did mention, however, that the genus *Bacillus* might prove an exception to that rule.[39]

In January 1876, the materialist John Tyndall entered into the Bastian controversy, claiming that all experiments reputed to support spontaneous generation were in error.[40] Six years earlier he had concluded that the power of air to develop life in sterile media paralleled its capacity to scatter light, and, on the strength of this, supported Pasteur's doctrine of limited panspermia: that germs were not uniformly distributed in the air.[41]

In his paper of 1876, Tyndall described numerous experiments using wooden chambers with a hinged rear door and two lateral glass windows. The top of the chamber carried a pipette and two narrow bent tubes, "so as to intercept and retain the particles" in the air, and the bottom was perforated by apertures, into which fitted large test tubes containing the experimental infusions.

Wherever the air in the chamber contained floating matter, as shown by the light beam passing through the windows, all the infusions putrefied. But in no case did such putrefaction occur when the air remained free of such matter. "The power of developing such life in atmospheric air, and the power of scattering of light, are thus proved to be indissolubly united."[42] These particles were assumed by analogy to be organic germs, it being "simply monstrous to conclude that they [the bacteria] had been spontaneously generated."

Despite such obvious betrayal of the "scientific method," Tyndall later had the gall to attack Bastian for being "unacquainted with the real basis of scientific inference" and even remarked that "it especially behoves me to take care that no theoretical learning shall taint my judgement of experimental evidence. I have always kept apart the speculative and the proved."[43]

It is difficult to understand why Tyndall was so opposed to the

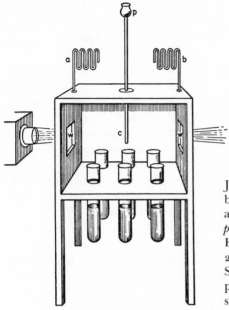

John Tyndall's wooden dust chamber. The diagram is from J. B. Conant, ed., *Harvard Case Histories in Experimental Science* (Cambridge, Mass.: Harvard University Press, 1957), vol. 2, in Case 7, "Pasteur's and Tyndall's Study of Spontaneous Generation," p. 526, and is used with the permission of the publisher.

doctrine of spontaneous generation, given his being both a materialist and an evolutionist. In his infamous Belfast Address of 1874, he had argued that one must either "open our doors freely to the conception of creative acts, or abandoning them, let us radically change our notions of matter." Matter, to Tyndall, was "the universal mother who brings forth all things, as the fruit of her own womb. . . . Believing as I do in the continuity of nature, I cannot stop abruptly where our microscopes cease to be of use. Here the vision of the mind authoritatively supplements the vision of the eye. . . . I cross the boundary of the experimental evidence and discern in that matter . . . the promise and potency of all terrestrial life."[44]

Despite such views and the experiments of Bastian and Sanderson, Tyndall concluded that, although one could believe in the possibility of spontaneous generation, one had to admit the "inability to point to any satisfactory experimental proof that life can be developed, save from demonstrable antecedent life."[45] One can only conclude that Tyndall's refusal to accept the evidence of the Bastian-Sanderson experiments resulted from his pragmatic adherence to the organismic germ theory of disease, which alone assured the "virtual triumph of the antiseptic system of surgery."[46] Tyndall thought that it was only through the complete elimination of the spontaneous generation viewpoint that medicine could hope to conquer disease. Tyndall was standing on a very insecure base at this time, and Bastian very quickly attacked his stand. Through the mouthpiece of an "Inquirer," he put three very succinct questions to Tyndall.[47] Does he accept Sanderson's experiments? If he rejects them, why does he do so? Is there another view which reconciles theory and fact? This "theory of reconciliation" was presented to Tyndall a few months later in a paper by Ferdinand Cohn.

In 1875, Cohn's second paper on bacteria had appeared, a summary of which appeared in the 29 June 1876 issue of *Nature*.[48] In this paper Cohn reported that the cheese bacillus, *Bacillus subtilis*, contained "oval or roundish, strongly refractive little bodies," which were probably spores able to resist boiling for a long period of time and subsequently capable of further development into bacillus rods. The article in *Nature* concluded with the statement: "One of Dr. Bastian's results is thus explained." In the same year Cohn began his investigations of hay infusions, a medium which had lent support to the views of Bastian. In his paper of July 1876,[49] Cohn reported the presence of *Bacillus subtilis* in hay infusions, which also had the power to produce heat-resistant endospores.

This paper was handed to Tyndall in the autumn of 1876. Tyndall immediately recognized that the results of the experiments of Bastian, Burdon-Sanderson, and Roberts, all of whom had reported the appearance of bacteria after boiling, could be explained by the presence of endospores, and that it was no longer necessary to assume experimental errors. In October 1876, Tyndall mentioned that the endospores could explain the latent and permanent existence of disease contagia.[50] And, in January 1877, he presented an address to the Royal Society, in which he admitted the difficulty of complete sterilization and also seemed to suggest that he had been less than honest in his previous work in which the appearance of bacteria had been denied. He remarked that, "cucumber infusion has been subjected, for intervals varying from 5 minutes to 5 ½ hours, to the boiling temperature without losing its power of developing life," and mentioned that, "precautions far greater than those found successful a year ago failed to protect these infusions from contamination."[51]

Bastian naturally reacted to this change of heart with justifiable scorn. Tyndall, he wrote later,

had completely stultified his previous position. He was no longer at issue with me and others in regard to the fact. The difference between us was now one of interpretation only. In spite of his previously much vaunted 500 negative results, and the good evidence which they supplied as to the death-point of Bacteria and their germs, Prof. Tyndall now endeavoured, as best he could, to cover his previous unfortunate position. The result was a complete change of front.[52]

The determination of "thermal death points" of bacteria, long a crucial issue in the spontaneous generation debate, now became even more significant. Unfortunately, the experimental determination of such temperatures remained impossible unless spontaneous generation itself was considered to be a chimera. If organisms appeared in a medium after boiling, opponents of spontaneous generation simply assumed them to have arisen from organisms or spores which had not been killed by the heat. That such results could equally well be interpreted as examples of spontaneous generation was dismissed by begging the whole question at issue: by claiming that spontaneous generation was impossible. On the other hand, Bastian had no experimental means of showing that superheating destroyed the "germinality of the fluids" rather than the enclosed spores. Once again the issue of spontaneous generation seemed unresolvable by experimental means. In order to extricate himself from this dilemma, it became necessary for Bastian to differentiate fluids which were capable of "generating" life from those which merely "nourished" it. Bastian claimed that by boiling a "generating" medium for ten minutes it was converted to a "nourishing" type, and by means of one such medium of boiled acid urine he set out to determine the "death point" of bacterial spores.

To one set of vessels containing acid urine he added a previously boiled fluid containing bacterial spores, and to another set a fluid containing bacteria lacking spores. Both sets were then plunged into a vessel of boiling water and allowed to remain there for ten minutes. After incubating both series at 50°C, he reported that, in each of 25 trials, "not one of either series had fermented. . . . Yet in control experiments with the same urine boiled for ten minutes in plugged vessels and subsequently inoculated with an unheated drop [of both fluids], fully developed fermentation was invariably set up from sixteen to twenty hours—showing clearly that there was nothing in the nature of the fluid to impede the development of the organisms."[53] In other words it appeared clear to Bastian that neither bacteria nor their spores could survive boiling temperatures.

The years 1876 and 1877 were key ones in the spontaneous generation debate. Not only did opponents of spontaneous generation claim that the existence of endospores rendered intelligible many of the previous incongruities, but the infamous *Bathybius* was reported to be a chemical artifact. As mentioned in chapter five, the idea of a universal protoplasmic *Urschleim* received popular support in Britain. None other than Ray Lankester had earlier supported the notion that abiogenesis was "a necessary and integral part of the universal evolution theory," remarking that, "a positive contradiction of the hypothesis of archigenesis is impossible." He asserted that a positive proof had now become possible, and that "Bathybius seems to be of the greatest significance for the theory of archigenesis. For if not through archigenesis, whence shall we derive this protoplasmic covering of the deepest sea bottom?"[54] But, in an account of the *Challenger* expedition, presented to the Royal Society in 1876, the chemist John Buchanan reported this "coagulated mucus" to be "sulphate of lime, which had been eliminated from the sea water . . . as an amorphous precipitate on the addition of spirit of wine."[55] Given the status of the protoplasmic theory at that time, it is not surprising that Huxley, who had received notification of the finding in a letter from Wyville Thomson dated 9 June 1875, should have accepted his own error. "Since I am mainly responsible for the mistake, if it be one, of introducing this singular substance into the list of living things, I think I shall err on the right side in attaching even greater weight than he [Thomson] does to the view which he suggests."[56]

Also in 1876, Bastian made another valiant attempt to justify his belief in spontaneous generation on experimental grounds. This was the famous urine debate, which climaxed with the strange confrontation between Bastian and the French Academy. Bastian opened the debate by stating that although boiled urine, which was acidic, did not usually ferment, it would do so if previously neutralized by liquified caustic potash or if maintained at a temperature above 38°C. He concluded that "neutralization increases the fermentability of urine," and that this was due either to a survival of germs or to the initiation of chemical changes by the potash that led to fermentation and the production of bacteria. He naturally opted for the latter explanation, on the grounds that boiled acid urine always remained clear until it was later neutralized by boiled caustic potash.[57]

In a second paper published one month later, he pointed out that the urine fermented only if neutralized exactly. Differing amounts of potash had to be added depending on the acidity of the urine, and if too little or too much were added the urine remained barren. Bastian argued that, were the potash a source of germs, the slightest drop would suffice to cause fermentation:

The answer is this: if boiled liquor potassae were a germ-containing medium, then one or two drops of it would always be capable of contaminating many ounces, or even a gallon or more of sterilized acid urine. This, however, is never the case. The boiled liquor potassae is only capable of imitating fermentative changes, and of leading to the appearance of bacteria when it is added in quantities strictly regulated by the quantity and degree of acidity of the specimens of urine with which the experiment is being made.[58]

The second paper was in reply to Pasteur's criticism that the boiled potash did indeed introduce germs into the urine. The issue was very clear cut to Bastian: "Can bacteria or their germs live in liquor potassae when it is raised to the boiling point? Such is now the simple issue to which certain great controversies have been reduced." The complete germ theory of Pasteur, according to Bastian, rested on the answer to this one simple question, for "if the germ theory of fertilization can be proved to be untrue, and if living ferments can be proved to originate spontaneously, we shall soon cease to hear much about an exclusive germ theory of disease." Instead the germ theory will be replaced by a "broader physico-chemical theory."[59]

Pasteur's claim was supported by Roberts and Tyndall, who both obtained negative results with the urine-potash experiment. However, they both knew the work of Cohn, and sterilization of both the urine and the potash was ensured by heating them to temperatures well above 100°C.[60] Negative results were also obtained by Pasteur and Joubert, who added pure solid potash to boiled acid urine. They confronted Bastian by echoing his previous remarks: "We have to do with a fact; yes or no; does urine which has been boiled so as to be sterile, and better still, fresh, natural urine, just from the bladder, not having been submitted to any preliminary boiling—does this at 50° yield organisms after being neutralized by potash?"[61]

Bastian criticized the paper of Pasteur and Joubert on the grounds that they had used solid rather than liquid potash and had used it in excess to produce an alkaline rather than a neutral medium. Bastian, convinced that sterile urine plus boiled liquid potash to neutralize would ferment, remarked, "I am perfectly ready to reproduce before competent witnesses the results of which I have above spoken."[62] The challenge was quickly taken up by Pasteur:

I defy Dr. Bastian to obtain, in presence of competent judges, the result to which I have referred, with sterile urine, on the sole condition that the solution of potash which he employs be pure, i.e. made with pure water and pure potash, both free from organic matter. If Dr. Bastian wishes to use a solution of impure potash, I freely authorize him to take any . . . on the sole condition that that solution shall be raised to 110° for twenty minutes or 130° for five minutes.[63]

Bastian's acceptance of the challenge led to one of the most bizarre episodes in the history of science.[64]

At the February 1877 meeting of the French Academy, Dumas, Milne-Edwards, and Boussingault were appointed to the Commission "to express an opinion on the fact which is under discussion between Dr. Bastian and M. Pasteur." That the Commission was called to pass judgement on a "fact" was significant in that Bastian held this to be the sole function of the Commission and held to this stand throughout the controversy.

The fact in question seems to me to be this—*Whether previously boiled urine, protected from contamination, can or cannot be made to ferment and swarm with certain organisms by the addition of some quantity of liquor potassae which has been heated to 110°C., for twenty minutes at least....* If the commission proposed to limit itself to reporting upon this mere question of fact I will willingly submit to its decision.

Bastian wrote to Dumas, the secretary of the Academy, on February 27, suggesting that both he and Pasteur perform their respective experiments before the members of the Commission and enquiring "exactly what steps the Commission proposes to take, and how the precise terms for formulating the question of fact which is to be submitted to their consideration are to be settled."

For some inexplicable reason Dumas did not reply to this letter until April 25, and this letter never reached Bastian. Bastian received his first delayed notification from Dumas on May 5, for the letter, again for some unaccountable reason, had been forwarded to the wrong address. Since this letter made reference to the earlier communication, which Bastian had never received, he wrote back to Dumas and again requested information "as to the precise question on which the Commission is to report," and "the mode in which the Commission will conduct the enquiry." Ten days later Bastian received a third letter from Dumas enclosing a copy of the missing letter of April 25. This letter read, in part: "The Commission, before entering into a full examination of the question, thought that it would be fitting to see at first the same experiments carried out freely by their authors." It was this threat to carry out a full investigation that concerned Bastian, who was insisting that the whole spontaneous generation controversy rested simply on the results of this single experiment. In a letter dated May 24, Bastian reiterated: "If it does not propose thus to restrict itself, and is empowered to express an opinion upon the interpretation of the fact attested, and on its bearings upon the 'Germ Theory of Fermentation,' or 'Spontaneous Generation,' then I must respectfully decline to take part in this wider inquiry."

Once more no French reply was forthcoming, so a month later Bastian wrote again to Dumas, reporting that he had made arrangements to visit Paris on or about July 15, but that "naturally before taking part in any arbitration I desire to receive some official intimation as to the exact terms and scope of the question which has been submitted to the arbitrators." A few days later Dumas replied: "It is desired, if it is possible, to concern ourselves only with your experiment and that of M. Pasteur on the subject of urine treated potash."

Bastian was still not satisfied, however, and only after a further letter from Dumas, dated July 12, did Bastian feel that the French had agreed to his limits. On July 15 he traveled to Paris and met with Dumas and Milne-Edwards. The latter, much to the annoyance of Bastian, immediately refused to participate in the Commission unless it had the power to demand addition experiments from the participants where deemed desirable.

The next day Bastian met with Dumas and proposed a compromise, namely that they should conduct "the first element" of the inquiry as defined by M. Dumas (i.e., simply repeating the experiments) and that "I should then return to London, and after the Commission had expressed its opinion . . . as to any variations in the experimental conditions which they might desire to institute, that I should return to Paris to witness and to perform such modified experiments." On July 18 Pasteur and Bastian presented themselves at the laboratory in the *École Normale.* What followed then is best expressed in the words of Bastian:

M. van Tieghem [the newly appointed replacement for Boussingault] was also there and shortly afterwards M. Milne-Edwards arrived. He apparently had had no communication with M. Dumas since the time of my interview, and when told, in reply to a question of his, of the proposition which I had made to M. Dumas, M. Milne-Edwards very hastily expressed his disapproval of it, and at once, without listening further, left the laboratory. He was followed by M. van Tieghem. I remained, and after one hour M. van Tieghem returned. He informed me that, having waited in vain for the arrival of M. Dumas, M. Milne-Edwards had at length gone away.

I remained in conversation with M. van Tieghem for nearly an hour in an upper room of M. Pasteur's laboratory. When we came down, much to my surprise, we learned from M. Pasteur that M. Dumas had arrived, that he had been told of the departure of M. Milne-Edwards, and that he also had then left, saying that the Commission was at an end. . . . Thus began and ended the proceedings of this remarkable Commission of the French Academy.

One could, I suppose, describe this event as another example of Anglo-French misunderstanding, for it is difficult to lay the blame on either side. One can certainly appreciate the position of Milne-Edwards, who obviously felt that such a complex question could not be settled in such a simple manner. One can appreciate his apprehension over the compromise situation, for had Bastian's experiments resulted in the appearance of bacteria it is doubtful whether he would ever have returned to Paris to carry out any modifications demanded by the Commission. Also, it is clear that a successful experiment would have been seen by Bastian as a denial of the "Germ Theory of Fermentation" and proof of spontaneous generation: the very questions he wished to avoid. On the other hand, the French were guilty of bad faith and appalling manners; they and Pasteur had agreed to

Bastian's terms before his arrival in Paris. The issue was further aggravated by the make-up of the Commission, which, similar to the one which had adjudicated the Pouchet-Pasteur controversy, did not contain, in the words of Bastian, "a single member who could be considered as representing my views, or even as holding a neutral position between me and my scientific opponents." The views of Milne-Edwards were well known and M. van Tieghem was a former pupil of Pasteur. One wonders why Bastian agreed in the first place to have his views tried by such a biased jury.

This episode was the virtual end of any claim that spontaneous generation could be shown to occur in the laboratory. In February 1877, Tyndall showed that complete sterilization could be brought about by a series of short boilings which gradually reduced the number of intact and resistant spores.[65] A year later he concluded: "The argument that bacteria and their germs being destroyed at 140°, must, if they appear after exposure to 212°, be spontaneously generated, is, I trust, silenced forever."[66]

This paper was one of three appearing in the 1878 issues of *The Nineteenth Century*, representing the final broadside exchanged between Bastian and Tyndall. Reviewing in detail his previous experiments in the wooden chambers, Tyndall noted, "that at least five hundred chances have been given to it, but it has nowhere appeared . . . in no instance was the least countenance lent to the notion that an infusion deprived by heat of its inherent life, and placed in contact with air cleansed of its visibly suspended matter, has any power whatever to generate life anew."[67]

From the fact that putrefaction occurred when air containing suspended particles was introduced into the chambers, Tyndall concluded "that the life we have observed springs from germs or organisms diffused through the atmosphere."

Bastian rightly denied the validity of Tyndall's conclusions. "The question really requiring to be solved," he remarked, "has always been whether mere organic *débris* from the air . . . could or could not also bring about such changes in suitable fluids."[68] Since Tyndall had never presented proof that his atmospheric dust contained bacteria or their germs, "all this discussion about the nature of the atmospheric dust, visible and invisible, together with elaborate and ingenious experimentation to prove its infective nature . . . has not really advanced the main question one iota. . . . We are no more able to say now than Schwann was in 1837 what is the precise nature of this something [in the air]."[69]

To Bastian, the basic question concerned the resistance of living things to boiling temperatures. Since he denied the existence of any form of heat-resistant protoplasm in the fluid state, he still regarded "the hypothesis of spontaneous generation as the most logical and consistent interpretation of the facts which are at present known."[70] At this stage, however, Bastian simply withdrew from the debate.

Tyndall's final paper, which involved a great deal of unjustified invective against both Bastian and Pouchet, concluded with reference to disease organisms: that, with spontaneous generation banished, no medical man need worry about the source of disease organisms. "Clearly assured that they are not spontaneously generated, his efforts will be directed to the discovery and the destruction of the germinal matter from which they spring. . . . This accomplished, the controversy comes to a natural end . . . life is too serious to be spent in hunting down in detail the Protean errors of Dr. Bastian."[71]

Despite the shortcomings of Tyndall's arguments, his work of 1876 to 1878 did see the virtual elimination of the belief that spontaneous generation could be shown to occur in the laboratory. Naturally, this had a marked impact on the question of the beginning of life. In the 1860s and early 1870s, it was possible and feasible for the evolutionists to extricate themselves from the "dilemma of abiogenesis," by claiming that the possibility of spontaneous generation had yet to be discounted. By the middle of the decade such a claim had become difficult to substantiate, since the results of Pouchet, Bastian, and others now seemed explicable through the work of Tyndall, Cohn, Koch, and the German cytologists.

The British, who had never been particularly enthusiastic about the link between evolution and spontaneous generation, quietly dropped the subject and one sees little further about it after 1874. In 1879, Huxley did again reiterate his earlier stand of 1860 that the denial of spontaneous generation "little affects the general doctrine [evolution]."[72] John Tyndall seems to have been one of the few British scientists to continue the debate into the late years of the decade. This was probably due to his adherence to two mutually contradictory beliefs, namely, materialism and opposition to any form of spontaneous generation. In his paper of 1878, which was directed toward Virchow's rejection of evolution, he acknowledged that: "The theory of evolution in its complete form involves the passage from matter which we now hold to be inorganic into organic matter; in other words, involves the assumption that at some period or other of the earth's history, there occurred spontaneous generation." Faced with the problem that spontaneous generation had been thoroughly discredited—in good part through his own labors—he concluded that this did not also discredit the theory of evolution, since its strength derived from "general harmony with scientific thought," rather than from direct experimental evidence.[73]

The Germans faced that greatest dilemma, since they had clearly linked evolution and spontaneous generation. Indeed, the materialists and Ernst Haeckel were enthusiastic about Darwin's theory precisely because spontaneous generation and the unification of the organic and inorganic into "one great fundamental conception," were implied. The materialists themselves were not particularly concerned with the problem posed by the discrediting of spontaneous generation, since abiogenesis was a metaphysical necessity requiring no proof. The less extreme physicalists, on the other hand, were much more aware of the dilemma. The position of du Bois-Reymond in 1872, that the origin of life was merely "an exceedingly difficult mechanical problem," no longer seemed viable. In 1876 he still adhered to spontaneous generation and wrote, rather desperately: "Be it that, in holding this theory [spontaneous generation], we experience the sensations of a man who, as his only hope of rescue from drowning, clambers on a plank which can only just keep him above the water. When the choice lies between a plank and drowning, the plank has a decided advantage."[74] He also attempted to claim some support for spontaneous generation by reference to the work of Johannes Müller, who had described the generation of molluscs from a special organ within the body of a holothuroid echinoderm![75]

One obvious escape from the dilemma lay in the theory of cosmozoa. In 1871 the great Helmholtz added his voice in support of this doctrine. "Who can say whether the comets and meteors which swarm everywhere throughout space, may not scatter germs of life wherever a new world had reached the stage in which it is a suitable dwelling place for organic beings."[76] Both Frederick Lange and Helmholtz argued that it was a perfectly correct scientific procedure to transfer any question into the realm of the transcendental when all attempts to generate organisms from inanimate matter had failed.[77]

To materialists, like Büchner, such an escape mechanism was unnecessary, abiogenesis being a perfectly feasible explanation. Others felt that the transfer was unacceptable, since it left the problem exactly where it was before. As August Weismann pointed out, "the mere shifting of the origin of life to some other far off world cannot in any way help us."[78] The assumption still remained that life only comes from preexisting life and leads thereby to the conclusion that life, like matter, is eternal.

Mechanists such as Weismann and Karl von Nägeli, who refused to avoid the issue by accepting the extraterrestrial origin of life, were then led to admit spontaneous generation as a "logical necessity" brought about by their commitment to the laws of physical causality. In the early years, Nägeli had allowed the possibility of spontaneous generation. In 1884, after the destruction of Schwann's theory of exogeneous cell development, and the work on nuclear division, Nägeli was forced to take the same stand as Weismann:

The origin of the organic from the the inorganic is, in the first place, not a question of experience and experiments, but a fact inferred from the law of the conservation of force and matter. If all things stand causally related in the material world, if all phenomena proceed along natural paths, then organisms which build up from and decay into the same stuff of which inorganic nature consists, must have originated primitively from inorganic compounds. To deny spontaneous generation is to proclaim a miracle.[79]

The link that Haeckel and others had made between the theory of evolution and the abiogenetic origin of life was so strong in Germany, that a vitalist was almost forced to reject Darwinian evolution from his abhorrence of abiogenesis. The rejection of spontaneous generation in the mid-1870s added strong support to this reaction. Such was certainly the case with Rudolph Virchow. It is fair to say that Virchow's reaction to Darwinism and to evolution in general was determined by his antipathy to the doctrine of spontaneous generation.

As has been mentioned in a previous chapter, Virchow was committed to a rejection of spontaneous generation. Not only did his metaphysical preconceptions of the nature of life preclude any notion of abiogenesis, but his scientific investigations had shown the falsity of adopting heterogenesis. Virchow seems to have accepted the transmutability of species as "a scientific necessity" before the work of Haeckel had transformed the doctrine of evolution into a monistic system in which an abiogenetic origin of life was a major part. Later, however, he reacted against the "crude schematization of the Darwinist," which seemed to him to be reminiscent of the old *Naturphilosophie*. His strongest statement on the issue was presented to the Munich meeting of the German Association of Naturalists and Physicians in 1877. There he clashed head-on with Haeckel over the teaching of evolution in the German education system. Virchow argued that "we should be abusing our power, we should be imperilling our power, unless in our teaching we restrict ourselves to this perfectly legitimate, this perfectly sage and unassailable domain."[80] Evolution could be taken only as a hypothesis, particular since the abiogenetic origin of life "lies at the basis of advanced Darwinism," and since science knows "not a single positive fact to prove that a *generatio aequivoca* has ever been made."[81] Finally, he presented a picture of the great dilemma:

No alternative remains when once we say, "I do not accept creation, but I *will* have an explanation." If that first thesis is laid down, you must go on to the second thesis and say, *"Ergo,* I assumed the *generatio aequivoca."* But of this we do not possess any actual proof . . . and whosoever supposes that it has occurred is contradicted by the naturalists, and not merely by the theologians. . . . Whoever recalls to mind the lamentable failure of all the attempts made very recently to discover a decided support for the *generatio aequivoca* in the lower forms of transition from the inorganic to the organic world, will feel it doubly serious to demand that this theory, so utterly discredited, should be in any way accepted as a basis of all our views on life.[82]

By 1880, even the Germans ceased to discuss the origin of life, although there were a few theories put forward during the first year or two of that decade. Most scientists in Britain and Germany took the stand that life either came from outer space or that abiogenesis had taken place only in the past. Since both concepts were beyond the scope of scientific investigation, the matter was thus no longer considered worthy of discussion. At that time, most scientists would have agreed with L. S. Beale, when he wrote:

Whether the generation of living matter was spontaneous or not cannot be proved, but much scientific speculation is built upon the theory of spontaneous generation. However necessary such a theory may be to the doctrine of evolution, there are no scientific facts which can at all warrant the conclusion that non-living matter only, under any conceivable circumstances, can be converted into living matter.[83]

At exactly the time that debates over the validity of experiments reputed to show spontaneous generation came to an end, it came to be realized that many diseases were indeed caused by specific bacteria. This theory stepped up the collapse of spontaneous generation, since only a few years previously proponents of spontaneous generation, such as Bastian, had argued that disease was a result of chemical processes in the body which could lead subsequently to the appearance of bacteria; that bacteria were a result, not a cause, of disease.

Pasteur himself drew attention to the relationship between disease theories and spontaneous generation when he wrote to Bastian in 1877. "Do you know why I desire so much to fight and conquer you?" he asked. "It is because you are one of the principal adepts of a medical doctrine which I believe fatal to progress in the art of healing—the doctrine of the spontaneity of all diseases."[84] Pasteur's belief that diseases were caused by specific living microbes followed naturally from his work on fermentation. In 1877 he entered the contagious disease controversy through his famous confrontation with anthrax, which had wreaked so much havoc in France. A year later, he, Joubert, and Charles Chamberland presented their famous address on the germ theory to the *Académie de Médicine,* in which they concluded that the cause of disease, "resides essentially and solely in the presence of microscopic organisms." We must, they argued:

for ever abandon the ideas of spontaneous virulence, ideas of contagious and infectious elements suddenly produced within the bodies of men or of animals giving rise to diseases.... All those opinions, which are engendered by the gratuitous hypothesis of the spontaneous generation of albuminoid-ferment material, of hemiorganisms, of archebiosis, and many other conceptions not founded on observation, are fatal to medical progress.[85]

At this time Robert Koch began his work on contagious diseases. His work between 1877 and 1884 on anthrax, tuberculosis, and cholera caught the imagination of the medical world. He claimed of his work on tuberculosis, for example, that: "It has been successful, for the first time, in furnishing complete proof of the parasitic nature of a *human* infectious disease."[86] Although the victory of the modern germ theory of disease was not so rapid and clear cut as Whig historians of medicine would have us believe, it is still clear that the impressive list of diseases which fell to medical bacteriologists during the later years of the nineteenth century, led gradually to the universal acceptance of bacteria as causal agents of disease. Although belief in miasmatic concepts, contagious particles of living matter, hemiorganisms, and the like collapsed, belief in the bacterial germ theory did not lead automatically to an acceptance of Koch's statement that: "If we grant that cholera depends on a well-defined specific organism, then we could not admit a spontaneous origin of cholera in any place. Such a specific organism . . . must always develop from its like, and cannot be produced from other things or out of nothing."[87]

Although all researchers agreed that disease-causing bacteria could not arise by a spontaneous generation from "nothing" or "other things," there was still a very widespread belief in pleomorphism. This tenet, which was particularly strong in Britain, denied the existence of distinct and stable bacterial species, and accepted that disease-causing bacteria could arise *de novo* from nonpathogenic strains under suitable conditions. As Karl von Nägeli expressed it, bacteria are

inconstant and continually lose themselves in one another.... The same organism in milk produces lactic acid, in meat, putrefaction . . . in the body to take part in some disease. As every communicable, infectious disease has arisen spontaneously it must continuously arise anew under similar circumstances. The miasmatic contagious diseases have an endemic distribution area where, with the simple cooperation of soil bacteria, they are produced anew.[88]

In other words, the contagious nature of disease was not necessarily implied from its bacterial cause. Disease could arise *de novo,* but no more by a spontaneous generation of disease entities.

Perhaps the most significant bacterial work pertaining to the doctrine of spontaneous generation was the use of a solid-culture medium by Robert Koch in 1881. Use of a fluid-culture medium had rendered the obtaining of pure cultures very difficult, since organisms could intermingle with each other. By spreading a thin film of bacteria over the surface of a cut potato, or by inoculating bacteria into a gelatin-nutrient mixture, and, by 1882, into an agar-agar medium, "every droplet or colony is a pure culture." Such a solid medium, wrote Koch, "is a solid soil, and prevents the different kinds of organisms (even those which are motile) from mixing with one another, whilst with regard to . . . the fluid substratum, there is no possibility of the different species remaining separate from one another."[89]

As a result of this technique, the source of any contamination of a culture medium could be determined, since any air-borne bacteria would result in the contamination of surface film of the agar only, and not the core of the jelly. In previous years, Pasteur's use of a fluid medium had prevented such a differentiation, thus rendering suspect any claims made regarding the source of contamination. With the introduction of agar plates, it became possible to distinguish bacteria arising as a result of inadequate sterilization techniques and those arising through aerial contamination. Therefore, it became easier for opponents of spontaneous generation to pinpoint faults in their opponents' experiments. This may well have been an additional factor which led to the rapid decline of any further experimental attempts to prove spontaneous generation in the 1880s.

This work on contagious disease and bacterial endospores also coincided with the discoveries of the German cytologists, which demonstrated the continuity of nuclei from cell to cell and from generation to generation. Between 1875 and 1878 more than one hundred and fifty papers were published by the German cytologists dealing with the problems of mitosis, reduction division, and fertilization.[90]

The concept that the fundamental basis of life could be ascribed to naked and simple protoplasm began to break down at the end of the 1860s. Numerous cytological studies had shown that the protoplasm contained organelles among which was the ubiquitous nucleus. Slowly the old definition of the cell returned. It was no longer a little mass of protoplasm but protoplasm containing a nucleus.

The role of the nucleus remained a mystery. By this time, that cells arose only from preexisting cells had become an observational fact, but the mechanisms involved were not understood. In other words, nobody could present a theoretical reason that simple organisms and cells could not be produced directly out of organic matter, any more than they could substantiate their acceptance of the fact that cells did indeed arise only and always from preexisting cells. One of the two mechanisms put forward to account for cell division was still that of endogeny, but since this seemed to be limited to plants, there appeared to be a fundamental distinction between plants and animals. Thereby the validity of Schwann's earlier claim that all organisms follow the same method of cell development was in question. In the animal kingdom the division of cells was thought to be preceded by the direct division of the nucleus; in plants the nucleus was seen to disappear and dissolve into the protoplasm. But even the zoologist Carl Gegenbaur admitted that animal cells arise endogenously. In 1874 he wrote that, in animal cells, "in many cases there appears to be a new formation of the nucleus."[91]

At this time new cytological techniques were being introduced which led to improved methods of fixing and staining. And, at the same time, the vexing optical problems of microscopy were being resolved with the manufacture of the apochromatic lens, subject to neither spherical nor chromatic aberration.[92] As a result, the various and complex bodies associated with nuclear division became visible and introduced further confusion into an already confused situation. Cytologists, such as Leopald Auerbach, Eduard Strasburger, and Francis Balfour, began to describe the appearance of the spindle and associated nuclear behavior. They all described the appearance of endogenous nuclei, however, particularly in the embryo sacs of plants, and all would have agreed with Balfour when he wrote that nuclei not only arise from simple division of preexisting nuclei, but also "by the complete solution of the old nucleus within the protoplasm of the mother cell, and the subsequent reaggregation of its matter to form the nuclei of the freshly formed daughter cells."[93]

It was the great German cytologist of Kiel, Walther Flemming, who finally described the process of indirect or mitotic division (*Karyokinetische Zelltheilung*). He described the appearance of the individual pieces of the chromatin at metaphase where they grouped on the equatorial plate, and the separation of the nucleic pieces into two daughter nuclei.[94] He showed further that the nucleic pieces were split longitudinally and each part subsequently distributed to the daughter nuclei. By 1882 Flemming generalized that all nuclei of vertebrate cells arise from preexisting nuclei by the process of mitosis,[95] and two years later the plant cytologist, Eduard Strasburger, finally came to accept the universality of mitosis in plant material.[96] Thus Schleiden and Schwann's theories of cell development were finally laid to rest. As pointed out in chapter four, these theories carried strong implications of spontaneous generation. In the words of Coleman: "The study of mitosis proved the continuity of cell and nucleus from the first cell body, the fertilized egg, to the final form of the organism."[97]

The most direct impact on the spontaneous generation controversy came from the discovery of nuclear continuity from generation to generation. Many cytologists, particularly Strasburger and his fellow botanists, maintained that the zygotic or "cleavage" nucleus arose endogenously immediately prior to fertilization. Following the work of Justus von Liebig, a chemical doctrine of fertilization prevailed, in which the sperm was seen to induce, by chemical agitation, the initiation of development processes in the cleavage nucleus. This "contact theory" of fertilization remained unchallenged until the 1870s. It seemed to justify the demand of very many biologists that vital processes must be explained by reference to physical and chemical phenomena, a demand which opposed the anatomically-based physiology of the past.[98]

In 1876 Oscar Hertwig published his famous paper on the fertilization and early development of a sea urchin egg, in which he concluded that the zygotic nucleus was produced by the direct fusion of male and female pronuclei.[99] The latter was thought to have been derived from a part of the egg nucleus that disintegrated immediately prior to fertilization, and the former was assumed, although not proven, to be derived from the entering sperm nucleus. Both Strasburger and Edouard van Beneden opposed this concept. "There cannot be," van Beneden wrote, "any genetic connection between the germinal vesicle [egg nucleus] or one of its parts, and the embryonic nucleus which appears in the egg after fecundation."[100]

According to van Beneden, the germinal vesicle disappeared during the prefertilization stages to leave a granulated and nonnucleated ovum. Fertilization was therefore seen as the fusion of the sperm nucleus with the yolk of the ovum. Following this, the two pronuclei arose endogenously from the zygotic protoplasm, one in the center of the cell and the other peripherally, immediately adjacent to the place of sperm fusion. The two pronuclei then came together to form the "cleavage nucleus."

In 1877 Hermann Fol described for the first time the complete continuity between the germinal vesicle and female pronucleus and the sperm and male pronucleus. He showed that the female pronucleus was derived directly from the germinal vesicle by the then-unknown process of reduction division or "retrogressive metamorphosis," as he also followed the entry and fusion of the sperm nucleus with the female pronucleus.[101]

It is clear that by the end of the 1870s the possibility of endogenous nuclei formation in the embryo sac of plants was still not completely denied. However, in the following few years work on chromosomes and particularly the establishment of the haploid-diploid features of the gamete and somatic nuclei led to the acceptance of complete nuclear continuity from cell to cell *and* from individual to individual. In 1893 Oscar Hertwig was still speaking of "so-called free cell formation," but the term no longer referred to endogeny but rather to the nucleus in a cell subdividing "several times consecutively, whilst the protoplasmic body remains undivided for considerable time without showing the least inclination towards even a partial cleavage."[102] He summed up: "Just as the cell is never spontaneously generated, but is produced directly by division of another cell, so also the nucleus is not created each time but always arises from the material of another nucleus. The dictum, *'Omnis cellula e cellula,'* is refined with the statement: *'Omnis nucleus e nucleo.'*"[103]

The doctrine of spontaneous generation seemed to have been finally erased. The mysteries of parasitic worms had long since been solved, and the appearance of infusorians and other microbes in previously boiled media seemed explicable by way of heat-resistant spores and cysts. And now, for the first time, work on mitosis, meiosis, and fertilization had established a basis for the continuity of life. As Edmund Wilson remarked in his monumental monograph on the cell:

The egg-cell, as well as the sperm-cell, arises by the division of a cell preexisting in the parent-body. *It is therefore derived by direct descent from an egg-cell of the foregoing generation, and so on ad infinitum.* Embryologists thus arrived at the conception so vividly set forth by Virchow in 1858 of an uninterrupted series of cell-divisions extending backward from existing plants and animals to that remote and unknown period when vital organization assumed its present form. Life is a continuous stream. The death of the individual involves no breach of continuity in the series of cell-divisions by which the life of the race flows onward.[104]

"Biological science today," wrote the English microbiologist and Wesleyan minister William Dallinger in 1877, "presents us with a magnificent generalization, and that which lies within it and forms the fibre of its fabric, is the establishment of a continuity—an unbroken chain of unity—running from the base to the apex of the entire organic series."[105]

But, as Bastian had stated many times and as was to be reiterated in the twentieth century, "Evolution implies continuity and uniformity," and they in turn "testify to the present occurrence of Archebiosis."[106] Despite the almost universal rejection of spontaneous generation, an undercurrent of doubt persisted—to appear again in the twentieth century.

EIGHT

DEMISE AND RECOVERY

INTO THE TWENTIETH CENTURY

1880–1930

The collapse of the doctrine of spontaneous generation was so complete by the 1880s that for the remainder of the century the scientific community remained virtually silent over the issue, as it also remained silent over the question of the beginning of life. In his presidential address of 1906 to the British Association for the Advancement of Science, E. Ray Lankester reminded his audience that protoplasm "holds in its meshes many and varied chemical bodies of great complexity," and that, "the question of spontaneous generation cannot be said to have been seriously revived within these 25 years."[1] Just as the physical sciences were underpinned by the ether concept and the atom theory, so the biological sciences rested on the theory of cellular protoplasm, the theory of evolution, and the concept of a specific organismic disease etiology. There was universal agreement that life could only be manifested at the level of the cell, viewed as a highly complex entity incapable of spontaneous generation from organic or inorganic matter. Bacteria, the lowest form of life, were regarded as "minute unicellular masses of protoplasm devoid of chlorophyll,"[2] similarly incapable of being generated spontaneously. When the American biologist C. O. Whitman, head of the biology department at the University of Chicago and director of the Woods Hole Marine Biological Laboratory, remarked in 1894 that:

We are no longer in the position of the philosophers of the last century, who were still totally blind to the central fact of modern biology—the law of genetic continuity, first neatly embodied in Virchow's formula, *Omnis cellula e cellula,* but since extended to every order of vital unit within the cell and raised to the full dignity of a general law by the final abandonment of the hypothesis of spontaneous generation,[3]

he was expressing a view with which few would have disagreed.

By 1880 the theory of evolution had become established in Britain and Germany, and even in France the first tentative steps towards a Lamarckian interpretation were being made.[4] As a result the problem of life's beginnings remained to be answered. It was difficult to claim that such an event or events took place only in the past, for such an event would be akin to a non-divine miracle; in the words of Henry Bastian, holding such a view was to "promulgate a notion which seems to involve an arbitrary infringement of the uniformity of nature."[5] William Preyer, in 1880, also criticized that view on the basis that, if conditions in the past had been so different from those of today that abiogenesis could have occurred, then the organisms must have perished immediately, since life can only be manifested over a very narrow range of environmental conditions. On the other hand, if conditions in the past had been no different from those of today, then abiogenesis must occur at present as well as in the past; "however, the failure of numerous attempts directed towards finding out how to do this has shown that it is unlikely to the highest degree."[6] One could evade the dilemma by continuing to believe in the eternity of life and its entry from outer space. Put forward seriously in the 1860s and 1870s by Lord Kelvin, von Helmholtz, Hermann Richter, Cohn, and Lange, in 1903 that hypothesis received additional support from the notable Swedish chemist Svante Arrhenius.[7] Although this concept had the advantage of denying spontaneous generation, while at the same time "asserting a consistent causal connection in nature,"[8] it is clear that most scientists chose to ignore the issue. They did so, as the English physiologist E. A. Schaefer put it, by "relegating its solution to some former condition of the earth's history, when, it is assumed, opportunities were accidentally favourable for the passage of inanimate matter into animate: such opportunities, it is also assumed, having never since recurred and being never likely to recur."[9]

This general disinterest in the issue of spontaneous generation is well illustrated by the negative response to the work of Henry Bastian. Having remained silent over the issue since his confrontation with Pasteur, Bastian suddenly, between 1904 and 1911, published a series of books reiterating his claims in support of spontaneous generation. Belief in the continual occurrence of abiogenesis was justified, he argued, "from the point of view of logic, of continuity, and of actual observation."[10] Quoting almost verbatim from his earlier writings, he argued once more that:

Evolution implies continuity and uniformity. It teaches us to look upon events of all kinds as the products of continuously operating causes—it recognises no sudden breaks or causeless stoppages in the sequence of natural phenomena. It equally implies that natural events do not vary spontaneously. It is a philosophy which deals with natural phenomena in their widest sense: it embraces both the present and the far-distant past. It assures us that the properties and tendencies now manifest in our surrounding world of things are in all respects similar to those which have existed in the past. Without a basis of this kind, the Evolution Hypothesis would be a mere idle dream. Uniformity is for it an all-pervading necessity.... But to assume, as the great majority of Evolutionists do that Archebiosis, or the natural origin of living matter, took place once only in the remote past and that it has not been repeated, or if repeated in past times, that it no longer goes on, is to look upon this process as a kind of natural miracle, and to postulate a break in continuity which ought only to be possible in the face of overwhelming evidence of its reality.[11]

Arguing that what one sees under the microscope is "a gradual emergence into the sphere of the visible, in some suitable fluid, of the minutest specks of living protoplasm," and that below that level of visibility "must be a process lying altogether outside human experience,"[12] he turned again to his experimental work on spontaneous generation. For Bastian, two problems remained to be solved:

We have (a) to ascertain as far as possible, by preliminary trials, the lowest amount of heat, and the duration of its application, which is necessary for the destruction of pre-existing living things within the experimental vessels. And we have (b) to see whether it is possible that fluids submitted to the *lowest necessary amount of heat* to ensure the destruction of all pre-existing life can still, by subsequent treatment under what are at the best only very unfavourable conditions, be induced to engender living matter.[13]

The impossibility of determining the "lowest necessary amount of heat" was now clear to Bastian. Any appearance of microscopic life in previously heated media was assumed by his opponents to imply the survival of living germs or spores, since to them spontaneous generation was an impossibility. On the other hand, the prolonged boiling necessary to destroy these germs was seen by Bastian as destroying the "germinality" of the fluids:

The last experiments I published on this subject, together with this kind of interpretation of their results given by others, were of such a nature as to convince me of the futility of further attempts in this direction. They tended to show the practical insolubility of the question by flask experiments with superheated fluids, because of the immediate renunciation by opponents of previous beliefs, whenever results would otherwise show that "spontaneous generation" had been demonstrated. For that reason I ceased to make and publish further experiments of this kind.[14]

After accusing his opponents of assuming rather than proving "various points needful to give adequate warranty to their interpretation," Bastian remarked:

They know quite well that spores are distinctly more resistant to heat than the parent organisms; they know that the parent organisms without spores are almost infinitely more common than such organisms with spores; they know that even under most favourable conditions such spores will often refuse to develop; yet whenever living organisms appear in the guarded fluids which have been heated far more than is necessary for the destruction of the parent organisms, they invariably assume that the much rarer spores have been present, and they further find it necessary to assume that the spores which are often most slow to develop under favourable conditions, now, in spite of the injurious heating and unfavourable conditions in which they are placed, straightway hasten to develop, grow and multiply. This is, surely, more like begging the question at issue, than judging in accordance with evidence.[15]

Bastian also attacked what he termed the "ultra-contagionist view" in regard to communicable diseases. Acceptance of spontaneous generation would, he stated, "pave the way for the admission that contagious diseases may also arise *de novo* instead of being only disseminated by contagia."[16] In this area Bastian stood on firmer ground; there were still those who believed that disease-causing bacteria arose *de novo* from bacteria of other types. As even Bastian admitted: "It was not even absolutely necessary that they should believe in a *de novo* origin of the Bacillus itself."[17] "There can, moreover, be little doubt that no impassable barrier exists between non-pathogenic and pathogenic Bacteria. The mutability in form, and changeability in activity of these micro-organisms is immense. They may merge into one another."[18]

Bastian had clearly forsaken his earlier chemical views on disease-causing agents, and while considerable evidence existed to substantiate his belief in the *de novo* origin of disease, few, if any, agreed with his statement with regard to leprosy:

It cannot be denied that the disease is to a slight extent contagious, but there is much evidence to show that in the main it arises *de novo*. There is certainly nothing unreasonable in the supposition that putrid fish, or other bad food of like kind, may carry into the system Bacteria and the toxic products which they have formed, and that the continued influence of such bodies may, in some persons, act as irritants, and either *engender* [italics added], or awaken organisms in this or that tissue having the characteristics of the leprosy Bacillus.[19]

Bastian had to admit what was obviously true. By the turn of the twentieth century, it was hopeless, "to convince persons, already firmly believing spontaneous generation to be a chimera, that any apparent success is not simply a case of "survival of germs."[20] One is reminded of the words of Max Planck, who wrote: "A new scientific truth does not become accepted by way of convincing and enlightening the opposition. Rather, the opposition dies out and the rising generation becomes well acquainted with the new truth from the start."[21] Indeed, this is what transpired. Just as in the physical sciences, where the ether concept and the atom theory collapsed, so in the biological sciences profound changes took place in the theory of cellular protoplasm and in the parasitic germ theory of disease. The result was a renewal of interest in the possibility of spontaneous generation.

Between 1880 and 1905 the prevailing view, that even the simplest living organism was extremely complex, opened up a hiatus between life and nonlife that seemed impossible to bridge by any fortuitous meeting of molecules. During the following twenty-five years, however, the gap was narrowed to such an extent that there appeared to be no discontinuity. With this came the renewed possibility of spontaneous generation, not of complex cellular organisms, but of subcellular "living matter." As Ben Moore, first professor of biochemistry at Liverpool University, remarked in 1921: "The territory of this spontaneous production of life lies not at the level of bacteria, or animalculae," rather it lies, "at a level of life lying deeper than anything the microscope can reveal, and possessing a lower unit than the living cell."[22] One may deny that this belief is akin to earlier concepts of spontaneous generation, in which fully formed organisms were deemed to appear. Both views are, however, essentially the same. Both accepted that an entity having all the essential characters of life could arise suddenly from matter that lacked these attributes.

Many different facets combined to reduce the distinction between life and nonlife, and thus to renew the claims of spontaneous generation. First, by the turn of the century, it was no longer feasible to believe that organic substances could only be produced by living organisms.[23] This old distinction between inorganic and organic matter had been slowly eroded away since Wöhler's work in the 1820s. This, therefore, allowed the substances from which "living matter" arose, to be quite complex. And, if the complex organic substances could arise abiogenetically, there was no longer need to make any distinction between heterogenesis and abiogenesis. Perhaps, the only doubts that persisted over the origin of organic compounds lay with their optical activity. In 1898, for example, F. R. Japp declared that "the *absolute origin* of the compounds of one-sided asymmetry to be found in the living world is a mystery as profound as the absolute origin of life itself," and that such molecules could not arise by any mechanical processes.[24] Attacked immediately by biometrician Karl Pearson and others, Japp concluded three months later that, "I no longer venture to speak of the inconceivability of any mechanical explanation of the production of single optically active compounds. ... I am as convinced as ever of the enormous improbability of any such production under chance conditions."[25] In a recent paper, George Kauffman has pointed out that the discovery by Alfred Werner in 1914 that noncarbon compounds also exist as mirror-image isomers removed "the last brick on the crumbling wall separating inorganic and organic chemistry."[26]

Belief that life may be manifested at a level below the cell may have been influenced by the discovery of subatomic structures. Bastian himself drew attention to this possibility, terming the electron "the infinitesimally minute primordial substance."[27] The American engineer William Brehmer, in a rather speculative article entitled "The Origin of the Living Organism in the Light of the New Physics," pointed out that, whereas in the past the atom was considered to be a "simple grain of matter," now it appears "it is really an organization of still simpler units which appear to be endowed with the most astounding energies." "The atom," he added, "is a dynamic organization subject to variation."

This dynamic concept of the atom is bound to have a far-reaching effect upon the entire gamut of human understanding. Among other things, we shall have to revise our present ideas regarding the nature of our environment. For if the very atom is a kind of elemental organism, does this not imply that the entire world about us is teeming with the essence of this something we call life? Can there really be such a thing as "dead" matter? Is this not rather an invention of the heretofore uninformed human mind? Is it going too far for us to wonder if, after all, the fundamental variable of the universe is not life?[28]

"This picturization of the living organism as a complex society of many kinds of energized particle-entities," he went on, "is most helpful to our understanding of the origin and the ultimate nature of the living organism."

Clearly, however, the strongest influence leading to a renewed interest in the possibility of spontaneous generation came from within the biological disciplines themselves. Here, the discovery of viruses and the rise of colloid biochemistry no longer compelled biochemists, "to consider living substance as possessing infinite complexity."[29] To a remarkable degree, the 1920s, which saw the climax of colloid studies in biochemistry and an enormous interest in viruses and bacteriophages, resembled the 1860s with its emphasis on anucleated globs of simple protoplasm. In both eras, the possibility of spontaneous generation was widely acclaimed.

In the late nineteenth and early years of the twentieth century, a series of plant and animal diseases was discovered whose agent was capable of passage through earthenware filters. The possibility of the existence of such agents was first suggested by Pasteur, but actual proof had to await the work of Dmitrij Ivanovskij (1892) and Martinus Beijerinck (1898), both of whom worked on the mosaic disease of tobacco. A year later Friedrich Loeffler and Paul Froschs' classic work on foot-and-mouth disease appeared, and by 1913 Benjamin Lipschuetz could list 41 animal and human diseases in which the filterable nature of the agent had been established.[30] These agents, variously termed invisible microbes, ultramicroscopic viruses, inframicrobes, or microplasms, gradually came to be called "filterable viruses." Although Beijerinck denied the corpuscular nature of these viruses, maintaining them to be a *"contagium vivum fluidum,"* improved filtration techniques led to the acceptance of their corpuscular nature. In these early years, most workers regarded them as minute animate beings, a view which reflected the paradigm of the "germ theory" of disease.

With the growing interest in viruses also came increasing emphasis on the colloidal nature of protoplasm. A comparison between molecular and colloidal solutions was first made by Thomas Graham in 1861.[31] Since proteins are approximately the same size as colloidal aggregates, many physical characteristics are shared. It was the parallel between certain properties of colloidal particles, or micellae, and living phenomena which led to analogies being drawn between living processes and micellae properties. The most important of these were surface phenomena (adsorption and the like) between the particle and liquid interface. Although Graham himself had drawn attention to the biological implications of colloid behavior, it was not until the 1890s that the colloid nature of proteins and protoplasm was first seriously studied.

In part, this change of emphasis reflected the inability of the light microscope to reveal any structure inside a cell which might explain the site of the many and detailed biochemical phenomena being elucidated at that time.[32] The colloid-liquid interface seemed to provide such a site. It also reflected early twentieth-century Positivist philosophy, stemming from Ernst Mach, who had striven to remove from science all hypothetical concepts, such as atoms and molecules. Colloid particles, on the other hand, were not hypothetical constructs but readily visible particles.

Another reason for this change of emphasis was the realization that the molecular weights of proteins were far in excess of anything previously measured. That a molecule was the smallest unit of a substance that retained the specific character of the substance lent weight to the belief that the measured proteins were not molecules in the strict sense, but "mixtures of substances whose composition is in fact much simpler than has been inferred hitherto."[33] In other words, functional proteins were not well defined macromolecules but "colloidal aggregates of indefinite composition."[34] Interestingly enough, many late nineteenth-century chemists regarded proteins as large molecules of definite structure. Crystallization of the hemoglobins and plant globulins had revealed their large molecular weights, and in 1920 and 1906 respectively, Franz Hofmeister and Emil Fischer put forward the polypeptide hypothesis of protein structure. Yet very quickly such views gave way to the new colloid approach to proteins, particularly following the work of Wolfgang Ostwald in Germany and Wilder Bancroft in the United States.[35]

But the most significant impact on thinking regarding the nature of proteins and protoplasm resulted from the rise of the new discipline, biochemistry. As Robert Kohler has recounted, between 1901 and 1905 the term "biochemistry" came into common usage and many biochemical journals appeared. Kohler argues that while the older "physiological chemistry," "was primarily concerned with the separation and analysis of animal material, biochemistry focused on the chemical changes of cellular metabolism."[36] Biochemists, stressing the dynamic nature of protoplasm, concerned themselves with enzyme studies and not with the study of cellular protoplasm. Kohler summarizes this change of outlook in the following terms:

Beginning about 1897 the protoplasm theory was replaced by a radically new theory of life processes, which ascribed each chemical change occurring in the living cell not to the whole protoplasm but to a specific intracellular enzyme. These specific protein catalysts, dissolved in the cell juice or attached to the structured protoplasm, carried out digestion, metabolism, respiration, and assimilation. Life, it was believed, was the self-regulating dynamic equilibrium of this system of catalytic reactions. This "enzyme theory," as it might be called, was the central dogma of the new biochemistry, and its acceptance coincides precisely with the emergence of biochemistry as a self-conscious movement.[37]

Pioneers of the new biochemistry, stressing the dynamic attributes of cellular activity, were mainly responsible for turning attention away from proteins of a definite static structure toward proteins of a colloidal nature. As early as 1901 Franz Hofmeister termed enzymes "colloidal catalysts," as did Gowland Hopkins in 1912.[38]

An early hint of this new approach to the simplest forms of life is to be found in Vernon Kellogg's influential book, *Darwinism Today*, published in 1907. Although attacking forcibly the belief that species could arise spontaneously, for which there was "absolutely no scientific evidence," he mentioned the colloidal character of protoplasm. Emphasis on the colloidal nature of protoplasm carried with it the assumption that protoplasm had an essentially simple physical structure, a view which, Kellogg admitted, was "very unsatisfying to most biologists," who "demand the assumption of an extremely complex structure."[39]

A year later, Archibald Macallum, professor of biochemistry at the University of Toronto, made the first definitive statement linking colloids, viruses, and spontaneous generation. Arguing that protoplasm consists of colloidal suspensions, each particle of which "is in a definite sense, alive," and which occur individually as viruses, he stated: "When we seek to explain the origin of life, we do not require to postulate a highly complex organism . . . as being the primal parent of all, but rather one which consists of a few molecules only and of such a size that it is beyond the limits of vision with the highest powers of the microscope."[40] In the past, he concluded, formation of amino acids and proteins took place countless times, until "one giving the right composition resulted in ultramicroscopic particles endowed with the chemical properties of ultramicroscopic organisms."[41]

The nature of this primordial living entity was a matter of conjecture. Not surprisingly, in this era of enzyme studies, some authors saw it as an autocatalytic protein enzyme. The Harvard biochemist Leonard Troland, for example, visualized the sudden appearance of such an enzyme in a system of slowly reacting substances. Under the influence of this enzyme the rates of such activities increased to surround the enzyme with an oily liquid. Such an oil droplet, he remarked, represents "the origin of the first and simplest life substance."[42] Later he termed this first autocatalytic particle "protase," similar to modern viruses and "chlamydozoans."[43] Others, such as Edward Minchin, reflected the growing interest in genetics, by arguing in favor of ultrascopic particles of the nature of chromatin. Terming them "biococci," Minchin, like Troland, saw present-day viruses and chlamydozoa as their closest relatives. Both writers saw these particles as the "starting point" of evolution, arising by chance and having all "the fundamental properties of living organisms in general."[44]

Discussion of viruses, colloids, and spontaneous generation reached its peak in the 1920s. At that time, a fierce debate took place over the nature of virus particles. Most of it revolved around Felix D'Hérelle's work on bacteriophages, which he initiated in 1917 with the discovery that stools of convalescent dysentery patients contained a filterable substance capable of lysing a culture of dysentery bacilli. He put forward three hypotheses to account for the nature of these filterable viruses:

1. Bacteriophagy may be *caused* by the presence of a foreign *chemical* principle, that is to say, a principle not derived from the bacterium which undergoes bacteriophagy.

2. Bacteriophagy may be effected through the action of a principle, either *chemical* or *living,* derived from the bacteria which undergo bacteriophagy.

3. Bacteriophagy may be caused by a *living* principle foreign to the bacterium.[45]

According to D'Hérelle, most workers supported either the third hypothesis or that version of the second which ascribed bacteriophagy to a normal autolysin of bacteria. He also remarked that, "Anglo-Saxon students favour the autolytic hypothesis, while those using the Latin languages, are with but few exceptions, in favour of the hypothesis of the living nature of the bacteriophage. The Germans are divided, some favouring the first, some the second concept."[46] This division is probably to be expected, given the authority of Pasteur and the "germ theory" in France, and the less enthusiastic acceptance of this theory in Britain. A more significant division, however, might well be found between chemists and the new breed of biochemists supporting the chemical theory, and bacteriologists and physicians supporting the "germ theory."

What is significant to the spontaneous generation issue, however, was the claim by D'Hérelle and others that, although the virus was a living parasite of bacteria, it was not a cellular organism. The study of bacteriophagy illustrated to D'Hérelle that "life does not require a cellular organization; that it results from a special physico-chemical state of matter, that is, the protein micella."[47] That there are "living micella beings or protobes" reflected, of course, the influence of colloid chemistry on biological thinking in the 1920s. "The micella is the smallest possible particle of matter in the colloidal state. The substance comprising all living beings is found in this state. Consequently the micella, the unit of colloidal matter is also the unit of living matter, and cells are constituted of a union of micellae."[48] That life can exist at the level of a colloid micella implied that no real barrier existed between life and nonlife. D'Hérelle himself was very explicit on this issue. "The fact," he wrote, "that there is such a being as *Protobios bacteriophagus* permits us to approach, in a scientific way, the problem of the origin of life."[49] He presented a phylogenetic tree, with the ultraviruses—"a group of unimicellar beings" at the base, and remarked that, "If we will ever see the time when we shall know the nature of the origin of life, it is certainly because the study of the ultraviruses, Protobiology, will reveal it to us."[50]

<pre>
                          Unimicellar Beings
                       (Ultraviruses—Protobes)
                    ╱   ╲

Plurimicellar beings   Plurimicellar beings
   (Bacteria)   (Undifferentiated Protozoa-Spirochetes)
      ‖   ‖
Cellular beings   Cellular beings
   (Fungi)   (Differentiated Protozoa)
      ‖   ‖
Pluricellular beings   Pluricellular beings
   (Vegetables)   (Animals)
</pre>

The "age of biocolloidology" also reached its climax in the 1920s, at which time journals abounded with articles dealing with the colloidal nature of protoplasm.[51] The significant aspect of these publications was the way in which some of the colloid chemists drew attention to the impact of their work on problems dealing with the beginning of life and spontaneous generation. Arthur Kendall, for example, in a paper read before the Second American Symposium on Colloid Chemistry, remarked that the distinction between primitive life and inorganic colloids was merely a "mental barrier," and that we could look forward to science producing a self-imbiding, dividing "bioidal complex."

A time comes when the statics of the chemical compound are merged into the dynamics of the colloid; mutable, plastic molecular complexes, in which the obstinacy of the chemical combination is replaced by a reactive, delicately poised equilibrium reminiscent in many of its properties of the phenomenon of primitive life.[52]

The American chemist Victor Vaughan, in his address to the United States Chemical Society in 1927, remarked: "The protein micella is the smallest possible particle of matter in the colloidal state. Possibly as D'Hérelle states it is the unit of living matter," and he went on, "we have evidence in the work of Du Novy that the cell may be but the logical consequence of the tendency of the protein molecule or molecules to establish dynamic equilibrium."[53] These chemists assumed that "life is fundamentally chemical," and that the equilibrium between a colloid particle and its environment is essentially akin to living processes. Given that assumption, then "there is no insuperable difficulty in the way of postulating some origin for living things."[54] Many references were made to E. C. Baly's production of formaldehyde and sugar from the action of ultraviolet light on carbon dioxide and water,[55] to proteins or glycoproteins as "living molecular structures," and to living entities as "a battery."

Erik Nordenskiold, in his famous *History of Biology* of 1928, rejoiced over the impact colloid chemistry had had on biology. Not only had it illuminated the structure of protoplasm, but also, he added rather mysteriously, "it has in many instances been possible to compare the mutual interpretation of the various structures even with physico-chemical metabolistic phenomena occurring in inanimate colloid substance."[56]

In 1928 an important series of articles dealing with colloids and viruses was drawn together by Jerome Alexander into a volume entitled *Colloid Chemistry*. All the papers stressed that life was "the expression of a particular dynamic equilibrium" and that the cell was a colloidal system. They stressed, too, that "the smallest amount of autonomous living matter is not the cell," and thus, with the demise of the cell theory, that the origin of life was no longer a mystery.[57] In the same vein, Charles Simon remarked that it was "scarcely conceivable, that the unquestionably wide gap between animate and inanimate matter should be barren of still lower forms than those with which biologists have thus far been familiar."[58] Alexander and Calvin Bridges considered *"a self-perpetuating and self-duplicating catalytic particle"* to be "the basic living units of all biological structures," whose simplicity is approached by bacteriophages, some viruses, and genes. In the past such particles, the moleculobiontia, had arisen directly from nonliving matter and by aggregation formed higher levels of complexity, the ultrabiontia (viruses and phages), bacteriobiontia, and, finally, the protobiontia.[59] Also in the 1920s, the famous American geneticist H. J. Muller described genes as "ultramicroscopic bodies," essentially similar to bacteriophages, and asserted that the first living material, "consisted of little else than the gene, or genes."[60]

The degree to which biochemists and geneticists had simplified the basic characteristics of life is nowhere better displayed than in a series of experiments designed to construct physical models having lifelike properties. Articles by two workers in this field, Stéphane Leduc, professor of medicine in Nantes, and A. L. Herrera, also appeared in Alexander's *Colloid Chemistry*. A. L. Herrera, a Mexican biochemist, seemed to have spent the better part of his active life investigating what he termed "plasmogeny," or research into the origin of protoplasm and the study of the lifelike behavior of artifacts. Some of his experiments seem to verge on the absurd:

Oleic, capric and caprylic acids and alkalis and alkaline carbonates, in water. A great variety of myelinoid forms, structures, movements, ciliary motion, striking resemblance to Protozoa and Protophyta.

Olive oil or rosin dissolved in gasoline. . . . Remarkable imitations of amoeba and infusoria—"Colpoids," artificial imperfect beings attacking and sucking themselves.

He denied that such artifacts were living things; rather, the experiments were said to "illustrate the physico-chemical concomitants of life."[61]

The assumption that life is explicable in terms of colloidal properties, which lay behind these early chemical papers on the origin of life, was fully shared by the young Russian biochemist Aleksandr Oparin. He made his first address on this topic to the Moscow Botanical Society in 1922, and his remarks appeared as a booklet in 1924. Although Oparin remained unknown in Western Europe and North America for another decade, this early work illustrates graphically the milieu in which scientists of all nations were discussing life and its origins in the 1920s. "There is no essential difference between the structure of coagula and that of protoplasm," he wrote.[62] Once colloidal solutions had come into being in the earth's watery covering, their molecules would gradually become more and more complex and would eventually gel. Life arose when the first gel came out of a colloidal solution, by chance:

The moment when the gel was precipitated or the first coagulum formed, marked an extremely important stage in the process of the spontaneous generation of life. At this moment . . . the transformation of organic compounds into an organic body took place. . . . With certain reservations we can even consider that first piece of organic slime which came into being on the Earth as being the first organism.[63]

Since each bit of slime was an individual that assimilated material at different rates, a "selection of the better organized bits of gel was always going on. . . . Thus, slowly but surely, from generation to generation, over many thousands of years, there took place an improvement of the physico-chemical structure of the gels." Naturally, as the amount of material in the surrounding medium diminished, "the more strongly and bitterly the struggle for existence was waged." Finally, only two courses of change were open to the gels, either toward a cannibalistic mode of life or toward autotrophism.[64]

A few years later, the English biochemist J. B. S. Haldane, who since 1923 had been working in Gowland Hopkin's laboratory at Cambridge and concentrating on enzyme studies, produced his famous paper on the origin of life. Echoing the views of biochemists of that era, he could claim that there was no absolute distinction between the living and the dead, but that rather, "since his [Pasteur's] death the gap between life and matter has been greatly narrowed."[65] Unlike Oparin, he particularly stressed D'Hérelle's work on bacteriophages:

"a step beyond the enzyme on the road to life, but it is perhaps an exaggeration to call it fully alive."[66] Like other biochemists he drew attention to the work of Baly, whose organic compounds produced by the action of ultraviolet light on water, carbon dioxide, and ammonia, "must have accumulated," in the past, "till the primitive oceans reached the consistency of hot dilute soup." The first precursors of life then obtained their energy from this soup by anaerobic fermentation. "Life may have remained in the virus stage for many millions of years before a suitable assemblage of elementary units were brought together in the first cell."[67]

At the conclusion of his paper, Haldane remarked that, although an acceptance of the abiogenetic origin of life was compatible with materialism, it was also compatible "with other philosophical tenets."[68] He may well have been thinking of his father, John Scott Haldane, who, from a very different philosophical position, was able to "adhere in the most thorough going manner to the conception of evolution,"[69] and thereby accept a natural origin of life. John Scott Haldane was utterly opposed to mechanical materialism, which he argued, "breaks down completely in connection with the phenomena of life."[70] From his work as an organismic physiologist, stemming from the tradition of Claude Bernard, he stressed the organism as "an organic whole," not as a sum of parts, and organic processes as "organically determined . . . in definite relation to the whole functional and structural activities of the organism." He believed life to be inherent in the so-called inorganic world and thus:

Evolution, therefore, takes on a very different significance. In tracing life back and back towards what appears at first to be the inorganic, we are not seeking to reduce the organic to the inorganic, but the inorganic to the organic. The apparently indefinite microscopical aggregations of formless colloid material which were at first taken for the origins of life from the inorganic, have gradually turned out to be definite living organisms. But biology will not stop at these, for they must have been evolved from something more primitive.[71]

Acceptance of a natural origin of life did not therefore necessarily imply a materialistic philosophy. Even more, it did not necessarily imply an acceptance of Marxist philosophy. This inference has been made by some, since two of the most influential writers of this period, Oparin and J. B. S. Haldane, became Marxists. This interpretation of the early work of Haldane and Oparin is certainly invalid, since dialectic materialism is opposed to the crude mechanical materialism of the 1920s.[72] Their work from this period rather reflects the impact of colloid chemistry, with its assumptions that colloid and living processes were essentially similar. Dialectic materialism, on the other hand, denied that living processes could be ascribed simply to physico-chemical interactions and thus that, at its simplest level, spontaneous generation was a feasible proposition. Indeed, as J. B. S. Haldane himself remarked in his *The Marxist Philosophy and the Sciences,* the views of J. S. Haldane were more in line with dialectic materialism than those of the crude materialists.

As usual, arguments ensued over whether or not spontaneous generation took place continuously or only in the past. Belief in the continuity of life had become so entrenched that most biologists either ignored the dilemma or attempted to explain why spontaneous generation no longer occurred. The usual explanation involved the ubiquitous bacteria which, it was claimed, destroyed any "living matter" produced by spontaneous generation. The absence of life in sterile flasks was not discussed! The existence of only one paleontological series was often raised to defend the existence of a single spontaneous generation in the past. J. B. S. Haldane argued for a single ancestor of all life by pointing out that all molecules of living beings display the same type of asymmetry, thereby rotating the plane of polarized light in the same direction. Since there seemed to be no reason "looking-glass organisms" should not exist also, he assumed that life arose only once or, "more probably, the descendants of the first living organisms rapidly evolved far enough to overwhelm any later competitors when these arrived on the scene."[73]

There were those, however, who supported the repeated and continuous spontaneous generation of simple life forms. A. Gardner[74] believed that viruses were being continuously generated as did Ben Moore:

There must exist a whole world of living creatures which the microscope has never shown us leading up to the bacteria and the protozoa. The brink of life lies ... away down amongst the colloids, and the beginning of life was not a fortuitous event occurring millions of years ago and never again repeated, but one which in its primordial stages keeps on repeating itself all the time and in our generation.[75]

In 1932, the renowned Belgian microbiologist André Gratia, who had become embroiled in a controversy with D'Hérelle over the original discoverer of bacteriophages, also questioned the "pasteurian theory" and suggested that precellular forms of life might continuously arise spontaneously.[76]

A. J. T. Janse, in an address to the South African Association for the Advancement of Science, supported a continuous spontaneous generation by subscribing to a remarkable Lamarckian interpretation:

If life arose only once in the past, how is it, he asked, that primitive organisms are still in existence today? "Either *all* Monera had the capacity to develop more or less rapidly into higher forms and *would thus eventually disappear as Monera,*" or they had not changed at all, in which case "evolution would be a dream." On the other hand, were a polyphyletic origin accepted, "all these separate 'trees' would have developed more or less along the same lines as took place with the very first, but [they] have only advanced to a certain stage, according to their date of origin." Thus, present primitive organisms "are descended from more recent primitive organisms formed *de novo* at different times, thus claiming that our present lowest forms are the most recent, instead of the most ancient."[77] By continuous abiogenesis such things as the geographical distribution of organisms and the presence of parasites seemed more readily explained.

All the views so far discussed agreed on one fundamental point. All agreed that life arose in a sudden chance act of spontaneous generation and that the entity produced by this act had all the essential properties of life. There was, naturally, disagreement over the nature of this first entity and what the essential properties of life actually were. This outlook was best expressed by Troland when he wrote:

Let us suppose that at a certain moment in earth-history, when the ocean waters are yet warm, there suddenly appears at a definite point in the oceanic body a small amount of a certain catalyzer or enzyme.... The original enzyme was the outcome of a chemical reaction, that is to say, it must have depended on the collision and combination of separate atoms or molecules, and it is a fact well known among physicists and chemists that the occurrence and specific nature of such collisions can be predicted only by the use of the so-called *laws of chance.* ... Consequently we are forced to say that the production of the original life enzyme was a chance event.... The striking fact that the enzymic theory of life's origin, as we have outlined it, necessitates the production of only a *single molecule* of the original catalyst, renders the objection of improbability almost absurd ... and when one of these enzymes first appeared, bare of all body, in the aboriginal seas it followed as a consequence of its characteristic regulative nature that the phenomenon of life came too.[78]

It was also well put by A. I. Oparin in his 1924 paper: "It is impossible, incredible, to suppose that in the course of many hundreds or even thousands of years during which the terrestrial globe existed, the conditions did not arise 'by chance' somewhere which would lead to the formation of a gel in a colloid solution."[79]

Such views clearly represent a continuation of earlier nineteenth-century views. In the 1920s, as in the 1860s, the distinction between inanimate matter and the most primitive life was so blurred that a fortuitous meeting of molecules to produce a simple living entity could be visualized. The only change lay in the nature of that simple living entity. The bacteria and simple anucleated blobs of protoplasm had been replaced by viruses, genes, autocatalytic enzymes, and similar particles of "living matter." The work of Pasteur, Tyndall, and others had had an impact, however. Despite the simple nature of these primitive living entities, most scientists were reluctant to admit that they could have been generated spontaneously in modern times. Precisely why this was not possible was never made clear. Some claimed conditions were not right, but such an explanation was meaningless—the only evidence of unsuitable conditions was the absence of such life.[80] In addition, if the simplest living entities were indeed mere aggregates of colloidal micellae, the argument claiming insufficient time to generate life in sterile flasks lacked credibility. Further confusion stemmed from an inability to define the basic nature of the most primitive living entities and from a lack of criteria by which a living thing could be recognized even if it had arisen in a flask.

By 1930, the issue of spontaneous generation was as confused as it had ever been. Despite the view that life could be manifested at a level far below the complexity of a cell, the problem of spontaneous generation rested in the same unsatisfactory position it had in the late nineteenth century. Just as in those earlier years, the issue could only be resolved by abandonment—by "relegating its solution to some former condition of the earth's history, when, it is assumed, opportunities were accidentally favourable for the passage of inanimate matter into animate."[81] Although some modern workers still adhere to such a view on spontaneous generation, most have abandoned such beliefs for reasons to be discussed in the final chapter.

NINE

FINAL ABANDONMENT

1936 TO THE PRESENT

In the opening chapter of John Keosian's *The Origin of Life*, published in 1964 and widely used by science undergraduates in North American universities, a distinction is made between mechanistic and materialistic hypotheses pertaining to the origin of life. The mechanistic, he wrote, "explains the origin of the first living thing in terms of the chance combination of the elements," while the materialistic, "takes a different approach in applying natural laws to the explanation of the origin of life." Instead of viewing the origin of life in terms of a living entity arising "all at once" by a sudden and chance combination, materialism, he argued, "views the origin of life as the result of a series of probable steps of increasing complexity, inevitably leading up to the living state."[1] From such a position it becomes completely meaningless, "to draw a line between two levels of organization and to designate all systems below as inanimate and all systems above as living."[2]

In that most modern biologists and biochemists accept that life emerged through a long and gradual process, belief in the possibility of spontaneous generation has been finally abandoned. Instead of arguing that life is analogous to a machine whose living characters suddenly appeared at the moment when all the pieces of machinery fell into place, they now argue in terms of an evolving process of gradually increasing complexity. "The pattern I propose," remarked John Bernal, the English physicist, in 1957, "is one of stages of increasing inner complexity, following one another in order of time, each one including in itself structures and processes evolved at the lower levels."[3]

Writings which stressed the evolution rather than the sudden appearance of life were not common in nineteenth-century literature. Herbert Spencer, of course, spoke in these terms as part of his all-encompassing "development hypothesis." To Spencer, "the affirmation of universal evolution is in itself the negation of an 'absolute commencement of anything.'"[4] Rather:

The advance from the simple to the complex, through a process of successive differentiations, is seen alike in the earliest changes of the Universe to which we can reason our way back; and in the earlier changes which we can deductively establish; it is seen in the geological and climatic evolution of the earth, and of every single organism on its surface; it is seen in the evolution of Humanity.... From the remotest past which science can fathom, up to the novelties of yesterday, that in which Progress essentially consists, is the transformation of the homogeneous into the heterogeneous.[5]

Spencer's views were not widely followed in the nineteenth century, however. In part this may have resulted from the controversy then waging over the age of the earth, which seemed to imply insufficient time for the evolution of living organisms to have occurred, let alone a long prebiotic evolution as well.[6] With the discovery of radioactivity by Henri Becquerel and the Curies at the end of the century, the restrictions imposed by physicists on geological time were gradually removed, and evolutionary concepts began to be put forward again.[7] In the eleventh edition of the Encyclopaedia Britannica, published in 1910, an article by Peter Mitchell suggested that because living organisms always arise from preexisting organisms, the terms "archebiosis" or "archegenesis" "should be reserved for the theory that protoplasm in the remote past has been developed from not-living matter by a series of steps."[8]

The idea that the appearance of life may have been preceded by a long chemical evolution grew out of early spectroscopical studies of stars and the realization by astronomers that different star types represent stages in a universal stellar evolution. Norman Lockyer, the famous English astronomer, even likened star types to the geological record. "We may," he wrote, "treat these stellar strata, so to speak, as the equivalent of the geological strata."[9] However, although they spoke of inorganic evolution preceding the appearance of life, the actual appearance was usually visualized as an "all-at-once" spontaneous generation. For example, E. A. Schaefer, in his famous 1912 address to the British Association for the Advancement of Science, spoke of the "process of evolution" being "universal" and seemed to deny the feasibility of spontaneous generation when he argued:

So far from expecting a sudden leap from an inorganic, or at least an unorganised, into an organic and organised condition, from an entirely inanimate substance to a completely animate state of being, should we not rather expect a gradual procession of changes from inorganic to organic matter, through stages of gradually increasing complexity until material which can be termed living is attained?[10]

Yet Schaefer did not attack the concept of spontaneous generation in general, merely that aspect of it that posited fully-formed complex cellular organisms, arising from inorganic matter. Indeed, he shared the then-commonly held view of biochemists that, "the chemistry and physics of the living organism are essentially the chemistry and physics of nitrogenous colloids," that the cell nucleus "possesses a chemical constitution of no very great complexity," and that, not only could the nucleus be prepared synthetically but also, with the chemical synthesis of a colloidal compound, "it will without doubt be found to exhibit the phenomena which we are in the habit of associating with the term life." Far from denying spontaneous generation, Schaefer concluded that "the possibility of the production of life—i.e., of living material—is not so remote as has been generally assumed."[11]

Later, in 1938, the American pharmacologist Reinhard Beutner stressed the unlikelihood of life being created in a flash, and spoke instead of "a very gradual development of the preparatory processes."[12] However, influenced by Wendell Stanley's crystallization of the tobacco mosaic virus in 1935, Beutner claimed that "a single molecule has the essential properties of a living organism,"[13] which at "some time and somehow, must have sprung from the inanimate matter left after this fiery mass [the Earth] had cooled down."[14]

Despite their references to the evolution of life, Schaefer and Beutner seemed to have had more in common with the "sudden" mechanistic hypothesis than with Keosian's "gradual," so-called materialistic hypothesis. They, like most other biochemists of that period, proposed that "the first living thing was a macromolecule, a 'living molecule,' that was formed by the chance coming together of the elements that composed it in the proper proportions and arrangements."[15]

The Liverpool biochemist, Ben Moore, put forward views much more akin to "materialistic hypotheses" than either Schaefer or Beutner. There was, he wrote in 1921, "a universal Law of Complexity," by which matter "tends to assume more and more complex forms." Thus, he argued, beginning with the ether, more and more complex inorganic structures arose as the temperature fell, eventually leading to the synthesis of carbon and the other elements. Carbon in turn formed complex molecules and colloids, properties of which, he noted, approximate those of living structures. Later, from such a matrix, autotrophic colloids formed; life had arisen. Although necessarily vague, Moore clearly did not visualize an all-at-once appearance of life. "It was no fortuitous combination of chances and no cosmic dust, which brought life to the womb of our ancient mother earth in the far distant Palaeozoic ages, but a well-regulated orderly development, which comes to every mother earth in the Universe in the maturity of her creation."[16] In addition, the process was conceived as a continuous one, primordial forms of life being generated "all the time and in our generation." Surprisingly he did not refute the claims of Bastian. Although more ready to accept the appearance of organic bodies than microorganisms in sterile flasks, he nevertheless felt that the latter may well appear "within a period of three to six months."[17]

What is so significant in Keosian's definition of the modern "materialistic hypotheses" concerning the origin of life, is the obvious reference to Marxist dialectical materialism in contrast to the "crude mechanistic materialism" of the nineteenth century. In particular, he writes of new laws operating at higher levels of organization which did not exist at the lower levels, and seems to take the Marxist position that denies the existence of events ascribed to what they see as lawless accidents of chance. "From the materialist view, the origin of life was no remote accident; it was the result of matter evolving to higher and higher levels through the inexorable working out at each level of its inherent potentialities to arrive at the next level."[18] In addition, Keosian's emphasis on the development of processes reflects an important aspect of dialectical materialism. That an American biochemist with no known Marxist affiliations should use such terminology under a heading of simple "materialism," reflects the enormous impact which Aleksandr Oparin has had on the issue of spontaneous generation and the origin of life. In 1936, Oparin produced perhaps the most significant book ever published on the problem. Entitled *The Origin of Life*, it was the first work to approach the issue from the standpoint of dialectical materialism.[19] In a very real sense, the acceptance of Oparin's basic position has led to the final abandonment of the spontaneous generation controversy.

Dialectical thinking with regard to nature—rather than to human society—is based on the writings of Friedrich Engels. Emphasizing the dynamic rather than the mechanical and static, Engels conceived of nature in terms of complex processes subject to a continuing historical development.[20] This development was explained by three universal laws of motion, all of which were incorporated into Oparin's writings: The Law of the Transformation of Quantity into Quality, The Law of the Unity and Conflicts of Opposites, and The Law of the Negation of the Negation.

The first law stands in opposition to biological reductionism as practiced in Engels' time. Such reductionism interpreted complex events in terms of the summation of the physical properties of its simplest parts. Thus reductionists tended to view life as a machine and the origin of life as analogous to the assembly of a machine. Dialectical materialism rejects these concepts: life is not a mere machine the origin of which can be explained by the chance combination of its component parts. Instead it postulates that, during the early history of our planet, a series of small quantitative changes took place, during which time qualitative changes also occurred. In other words, as the developmental processes slowly progressed from atoms through more and more complex molecules to simple cells, "the old laws of physics and chemistry naturally continued to operate, but now they were supplemented by new and more complicated biological laws which had not operated before,"[21] and which are not deducible from the physical laws. In the words of Engels himself, the protein "is something essentially different from the molecule, just as the latter is different from the atom."[22] Implicit in this belief lies the concept of causality. To many Marxist writers, any abandonment of rigid determinism in favor of statistical laws seemed to undermine this concept by appearing to postulate effects with no cause. Hence, for example, the well-known Marxist objections to quantum physics and Werner Heisenberg's uncertainty principle and Oparin's continued hostility to notions that life arose "by chance."

The second law, the law of conflicts, sees motion as stemming from "conflicts" of internal elements which can only exist in relation to one another. Hence Oparin's emphasis on the conflicting metabolic processes of anabolism and catabolism as the basis for genetical change, rather than the static DNA template. Finally, according to the third law, the emergence of new qualities as a consequence of quantitative changes, implies the "negation" of the previous quality which thereby may prevent the appearance of this quality again. Oparin's assumption that the very existence of life on this planet negates any further emergence of it, reflects this law.

According to Loren Graham, by 1936 Oparin had shown a marked shift towards Marxist interpretations of the origin of life.[23] Basically, the Oparin of 1924 was a reductionist who believed life, which had arisen suddenly by chance, was completely explicable in terms of physics and chemistry. His shift away from this approach may have reflected in part the changed political climate of the Soviet Union. In the 1920s, the intellectual scene was relaxed, with no attempt being made to impose ideological conformity on the scientific community. The scientific institutions remained in the hands of prerevolutionaries, and as late as 1929 no member of the Academy of Sciences was a member of the Communist Party. Between 1927 and 1929, however, Stalin launched his massive agricultural, industrial and cultural revolution during which control of scientific institutions passed into the hands of the Communist Party. As a result, a tendency developed to politicize science, to associate prerevolutionary science with bourgeois thinking. Oparin himself, in the opening paragraph of his 1936 work, refers to arguments over the origin of life as a reflection of "the underlying struggle of social classes."[24] At this time too, there developed an emphasis on utility in opposition to the theoretical nature of science practiced by the Western trained prerevolutionary scientific elite. In 1935, for example, Oparin published a work dealing with the biochemical base of tea production. Also, in 1929, the agronomist Trofim Lysenko announced that wheat, normally planted in the winter, would ripen after a spring planting if subjected to moisture and low temperatures immediately before sowing. This process of "vernalization" obviously had profound practical implications in a country which suffered severe grain mortality during the winter months. In comparison, Western genetics, with its emphasis on fruit flies and wrinkled peas, seemed utterly sterile and useless. Not surprisingly Lysenko became a hero of socialist agriculture, and a special laboratory was built for him at the Ukrainian Institute of Selection and Genetics in Odessa. By 1935 over five million acres of the Soviet Union were being planted with vernalized winter cereals.

Concomitant with these events, many Soviet scientists began to reconstruct their science from the viewpoint of dialectical materialism. Lysenko in 1935 began using the rhetoric of dialectical materialism in his support of progressive, innovative, Darwinian, Marxist genetics, and in his attacks on clerical, antiscientific, and "methodologically bourgeois" science. Loren Graham argues that this growing tendency among Soviet intellectuals in the 1930s to use the language of dialectical materialism was not simply a political and rhetorical device to gain favor. Rather, these intellectuals, among whom was Oparin, "found historical and dialectical materialist explanations of nature to be persuasive on conceptual grounds."[25] Oparin, in his 1936 text and all subsequent writings on the problem, used the dialectical framework in a very persuasive and successful manner to attack the problem of the origin of life. It is unfortunate that most Westerners associate Marxist biology with Lysenkoism and have failed to realize some of its very positive results.

By the 1930s, with the rapid decline in colloid biochemistry and the emphasis on proteins as definitely structured "macromolecules," the innate complexity of living entities was once again being stressed.[26] All living entities, whether bacteria or viruses, must, in the words of Oparin, "be endowed with a definite and complex organization which makes it possible for them to perform a number of vital functions."[27] As such, it was inconceivable that they "could appear in a very short time, before our eyes, so to speak, from unorganized solutions of organic substances."[28] Thus, he argued, since life cannot arise spontaneously, "it must have resulted from a long evolution of matter, its origin being merely one step in the course of its historical development,"[29] and in the course of this evolution "more and more complex phenomena of a higher order became superimposed upon the simplest physical and chemical processes."[30] Eventually, there emerged systems subject to biological laws.

Oparin then set out the various stages in the evolutionary process. In doing so, he emphasized the abiogenetic formation of organic substances and the necessity of a sterile, lifeless planet on which these early stages could occur. Without such a prerequisite these early organic substances would have been destroyed by microorganisms. Of course, in the 1930s it was no longer believed that organic substances were producible only by living organisms, and thus, that the most primitive life must have arisen from inorganic matter and have been autotrophic, that is, capable of manufacturing organic substances from inorganic material.

Oparin described the first stage in the long evolutionary process as the appearance of a primary reducing atmosphere, in which carbon existed as hydrocarbons and cyanogen, and in which nitrogen occurred as ammonia, a view that was then commonly held by astronomers.[31] The second stage saw the production of more and more complex organic substances from the initial hydrocarbons, which "are pregnant with tremendous chemical possibilities."[32] As a result of condensation, polymerization and oxidation-reduction reactions in the warm waters of the primary seas, Oparin suggested that "numerous high molecular compounds, similar to those present in living cells, may appear," and that, "there is absolutely no reason to doubt that these reactions were essentially like those chemical interactions which can be reproduced at the present time in our laboratories."[33] Such reactions included the production of amino acids and other protein-type compounds.

The effect of dialectical thinking appears most clearly in Oparin's discussion of the third stage in the evolutionary process, a stage he called "the origin of primary colloidal systems." Instead of seeking to define living properties in terms of the structure of organic molecules, he argued that "the laws of organic chemistry cannot account for those phenomena of a higher order which are encountered in the study of living cells."[34]

Polymerization of pure organic molecules will not result in the appearance of living processes, and neither are the properties of mixtures of such substances merely "the sum of the properties of their components." Rather, he argued, "on mixing different substances new properties appear which were absent in the component parts of the mixture." "This alone compels us, in considering the evolution of organic substance, to rely not upon those alterations to which this or another isolated compound may be subjected, but to bear in mind alterations which take place in complex mixtures of various organic substances."[35]

Oparin discussed at length one such reaction which takes place in mixtures of organic substances, namely the self-formation of polymolecular open systems or "coacervates." The word was coined by the Dutch chemist Bungenberg de Jong in 1929 to describe the appearance of two liquid phases in hydrophilic colloids: the coacervate, rich in colloidal substances, and a noncolloidal equilibrium liquid.[36] Although many properties of such coacervates may be compared to those of protoplasm, Oparin cautioned that they should not be regarded as models of protoplasm but merely as an "essential landmark in the evolution of our appreciation of the physico-chemical properties of the protoplasm."[37] Among such properties, Oparin mentioned increase in size due to adsorption of various substances in the equilibrium liquid and the appearance of a definite structure within the coacervate droplets.

What is particularly important in this work of Oparin was his denial of the appearance of "a single, definite, individual substance," which had characterized previous discussions of the origin of life. In its place, Oparin conceived of "a complex mixture of different high-molecular organic compounds of primary proteins, lipids, carbohydrates and hydrophil colloids," the formation of which was "unavoidable."[38] With coacervate formation, "organic matter became concentrated at different points of the aqueous medium and, at the same time, sharp division occurred between the medium and the coacervate."[39] "To initiate life," he concluded, it was necessary that these coacervates "acquire properties of a yet higher order, properties subject to biological laws."[40]

The biological laws of natural selection came into play, Oparin argued, as a result of differing chemical reactions taking place in the coacervates. Some would become unstable and disappear; others would gradually acquire minute amounts of enzyme and thereby gain a selective advantage. In time, therefore, their "inner chemical organization became strengthened in the process of natural selection, insuring a gradual evolution which finally culminated in those highly perfected enzyme systems existing at the present time."[41] This new biological factor of natural selection

raised the colloidal systems to a more advanced state of evolution. In addition to the already existing compounds, combinations and structures, new systems of coordination of chemical processes appeared, new inner mechanisms came into existence which made possible such transformations of matter and of energy which hitherto were entirely unthinkable. Thus systems of a still higher order, the simplest primary organisms, have emerged.[42]

The impact of Oparin's work was profound. Although not the first writer to view the beginning of life as a gradual process, he was the first to set out a series of hypothetical steps by which life might have emerged. That it would be possible to duplicate these steps in the laboratory meant that the problem of life's beginnings became a legitimate and fruitful scientific research problem. In place of searching for a fortuitous meeting of molecules to produce an indefinable and unrecognizable living entity, biochemists could now investigate each of Oparin's stages. For this reason alone it had appeal beyond the Soviet Union, where the Marxist interpretation proposed by Oparin provided an added source of support.

The antimechanist approach of Marxist biologists such as Oparin paralleled a similar movement in Europe and North America. As Garland Allen has recently shown, during the 1920s and 1930s a trend developed in biology away from mechanistic materialism and toward a more holistic, organismic viewpoint. Stemming in part from the new quantum and relativity physics, this movement was reflected in the organismic physiology of Charles Sherrington, Walter Cannon, Lawrence Henderson, and others, in opposition to the classical gene concept; and in the embryological work of Hans Spemann, Paul Weiss, and Ludwig von Bertalanffy. These scientists displayed an intense interest in biological *interactions* and the environment in which these interactions took place. Like their Marxist counterparts, they also argued that the whole is greater than the sum of the parts. Both groups felt that knowledge of individual chemical reactions within the body was insufficient in attempting to understand the functioning of the complete organism, and that a greater emphasis on and awareness of the relationships among these chemical reactions were required. From this movement concepts such as homeostasis, intergrative action, self-regulation, and negative feedback appeared in physiology, while in embryology Paul Weiss was developing his field theory.[43] The widespread acceptance of Oparin's work can be understood when seen against this generally organismic milieu of biology, although, clearly, within the narrow confines of biochemists interested in the origin of life problem, it was the experimental opportunities flowing from the Oparin approach that had the most direct appeal.

With the end of World War II, biochemists began to investigate each of the stages proposed by Oparin. The most famous among them was Stanley Miller, who in 1953 synthesized amino acids from an "atmosphere" of methane, ammonia, water, and hydrogen. The bearing of this work on Oparin's thesis was made explicit in Miller's opening remarks in the article in which he pointed out that such an atmosphere "was suggested by Oparin."[44] In a sense, this synthesis differed little from nineteenth-century productions of organic substances. Had such a synthesis been possible at that time, only the materialists would have seen it in terms of the origin of life question. To most chemists and biologists of the nineteenth century a clear distinction existed between organic matter and living organisms. After Oparin, however, any production of organic material was viewed as a step on the long evolutionary road, particularly when the ingredients from which the amino acids arose seemed similar to the reducing atmosphere presumed to exist on the prebiotic earth.

In 1954, George Wald published an important article on the origin of life in *Scientific American,* in which he stated that Oparin's book "provides the foundation upon which all of us who are interested in this subject have built."[45] He pointed out that many organic molecules show "a spontaneous impulse toward structive formation," and gave as an example the precipitation of structured collagen fibrils from a free solution with completely random orientation. This self-assembly of collagen fibrils was the subject of a paper by F. Schmitt, published in 1956.[46]

Naturally, not all scientists were in agreement with Oparin's views, and in the 1950s two antithetical approaches to the problem came into sharp focus. There were those who, with Oparin, conceived of a long evolutionary development of life, and there were those who still spoke of a sudden, very improbable event leading to the instant formation of the first living molecule. Such a spontaneous generation of life was inevitable, "given sufficient time, and sufficient matter of suitable composition in a suitable state."[47]

The conflict between these two points of view in the 1950s must be seen against the backdrop of international politics. It was the time of the cold war, when dissenters in the Soviet Union and the United States were subjected to intolerable political and personal harassment. There was a cold war, too, in science, which impinged directly on the origin of life controversy. In 1948, the Soviet Communist Party had prohibited further teaching and research in classical Mendelian genetics as part of its ideological support of Lysenkoism. Classical Mendelian genetics had, of course, become established in the Western world following the famous *Drosophila* studies, which Thomas Hunt Morgan and colleagues at Columbia University began in 1908. It was, in addition, one of the first really significant contributions the United States had made to world science. Naturally, therefore, Americans took a legitimate but nationalistic pride in bolstering their scientific arguments in support of Mendel and Morgan. Professor Ralph Spitzer lost his job at Oregon State University in these years, for daring to suggest that Lysenko's work ought to be examined.

Three factors converged at this time to bring the issue of the origin of life into the political arena. First, Oparin and many of his supporters were obviously Marxists using the rhetoric of dialectical materialism[48]; second, Oparin himself was closely associated with Lysenko; and third, the major opposition to Oparin came from the American geneticists and those working on bacteriophages who, after the work of James Watson and Francis Crick, conceived of the origin of life in terms of a sudden spontaneous generation of the first DNA molecule.

In 1956, a revised and much enlarged edition of Oparin's 1936 work appeared. Entitled *Origin of Life on the Earth* and translated into English a year later, it presented dialectical arguments more forcibly and clearly than before. In the introductory chapter, for example, after criticizing any attempt "to explain the origin of life by separating it from the general development of matter," Oparin wrote:

A completely different prospect opens out before us if we try to approach a solution to the problem dialectically rather than metaphysically, on the basis of a study of the successive changes in matter which preceded the appearance of life and led to its emergence. Matter never remains at rest, it is constantly moving and developing. . . . Life thus appears as a particular very complicated form of the motion of matter, arising as a new property at a definite stage in the general development of matter.[49]

Oparin was deeply involved in the Lysenko controversy. Zhores Medvedev reports that, in 1948, while heading the biological section of the Academy of Science, Oparin refused to allow the appointment of plant physiologist D. A. Sabinin, an opponent of Lysenko, to a post in the soils section of the Academy.[50] In 1950, Oparin joined with Lysenko in supporting the award of the Stalin Prize to Olga Lepeshinskaia for her work on the spontaneous generation of cells from a noncellular nutrient medium. Graham describes this episode as "one of the low points" at Oparin's career, for, both before and after this, he was opposed to any such sudden appearance of fully-formed cells. He later came under increasing criticism from Lepeshinskaia and her supporters.[51] Finally, in 1955, when Lysenkoism had lost much of its support among Soviet scientists, a petition, signed by more than three hundred of them, requesting Lysenko's removal from the presidency of the Lenin All-Union Academy of Agricultural Sciences and Oparin from the secretaryship of the Academy of Science, biological section, was granted.

That life originated with the appearance of the gene was first stated by the American geneticist H. J. Muller in 1926.[52] Attracted by his communist sympathies, he visited the Soviet Union in 1933, only later to become an opponent of both Stalinism and Lysenkoism. In direct contrast to Oparin, he continued to believe that life could be so well defined that the exact point at which it began could be determined. In other words, he accepted that life began by a spontaneous generation. In 1947, criticizing the "so-called organism-as-a-whole" view of life, whose origin was "an incomprehensible enigma," he argued that, "all other material in the organism is made subsidiary to the genetic material, and the origin of life is identified with the origin of this material by chance chemical combination."[53] Muller restated these views in 1955, two years after Watson and Crick had published their model of DNA structure.[54] To Muller and those who agreed with him, the gene is the unique "living" molecule, which is characterized by the ability to reproduce itself and mutate and to produce specific metabolic enzymes.

Such a view evoked strong reactions from Soviet scientists during the first International Symposium on the Origin of Life, which took place in Moscow in August 1957. Although Oparin claimed correctly that "the principle of the evolutionary origin of life" provided a basis for the program of the symposium, he erred in stating further that "this principle was shared by all the participants in the conference."[55] It was not shared by a number of American visitors such as Norman Horowitz, Wendell Stanley, and Heinz Fraenkel-Conrat, although Sidney Fox, Erwin Chargaff, and Stanley Miller clearly shared it.

Wendell Stanley, in a paper entitled "On the Nature of Viruses, Genes and Life," argued that viruses and genes were basically nucleic acids and thus, "the distinction between living and non-living things . . . seemed to be tottering." He further maintained that with nucleic acids, "we are dealing with life itself."[56] Horowitz took a similar view. Arguing against Norman Pirie's well-known claim that life is indefinable, he claimed life was manifested in mutability, self-duplication, and heterocatalysis and that it "arose as individual molecules in a polymolecular environment."[57] The same position was taken also in a paper by Fraenkel-Conrat and B. Singer.[58]

Erwin Chargaff, in his usual forceful way, disagreed with his countrymen. "Is life itself only an intricate chain of templates and catalysts and products?" he asked. "Is the cell really nothing but a system of ingenious stamping presses, stencilling its way from life to death?" "No", he answered, "for I believe that our science has become too mechanomorphic."[59]

The Marxists present naturally took issue with the "mechanomorphic" point of view. Bernal, for example, after arguing that nucleic acids' arising by a fortuitous combination of elements would have taken longer than the life of the universe, claimed that, "the problem has been wrongly posed," and continued: "There is no question, to anyone who has examined the evidence, of the need to explain the origin of life as consisting of one decisive step, because it plainly did not originate as such."[60] Rather than using the easily misconstrued term "spontaneous generation" to describe the appearance of life, Bernal introduced the term "biopoesis" to the meeting, to include all the slowly evolving, life-making processes.[61]

The Soviet Marxists, who were present in large numbers, also attacked those who claimed DNA to be a "living molecule." Aleksandr Braunshtein, for example, who in 1959 became laboratory chief at the U.S.S.R. Academy of Sciences Institute of Radiation and Physico-Chemical Biology, stressed again that "the transition to life only occurs when these compounds and many other substances are unified into complicated systems."[62] Some of the participants were known supporters of Lysenko, who, by 1957, had regained his hold on Soviet genetics with backing from Nikita Khrushchev. One such supporter was Nikolai Nuzhdin, a geneticist in the Institute of Genetics at the U.S.S.R. Academy of Sciences, who had already published papers attacking what he termed "reactionary Mendelism-Morganism." He saw in Stanley's paper "the evergrowing tendency to ignore the qualitative specificity of living material which distinguishes it from nonliving material." Such a tendency had entered biology, he claimed, from the work of physicists, who "consider the possibility of a more complete explanation of biological phenomena solely in terms of their understanding of the laws of physics and chemistry." Mentioning next the genetic concepts of the horticulturist Ivan Michurin, who opposed theories involving specific carriers of heredity, he concluded:

I am glad to point out that, both in Academician Oparin's monograph and in many of the papers, many suggestions have been made concerning the question on which I am touching. It will undoubtedly play a great part, not only in the study of the origin of life, but also in genetics. From what has been said it clearly follows that, however important the part played by nucleic acid in biological processes, its molecule was not the original basis of life. It is not a "living molecule" but one of the parts of a living structure which can only fulfil its biological function against the general background of metabolic processes taking place within the cell. The synthesis of the protein molecule depends on nucleic acid just as much as the synthesis of the latter depends on the protein molecule. This is quite understandable. Living material is a complex of compounds which determine the regular course of the processes which constitute life. One cannot isolate an individual component without interfering with the living material, still less can one understand life on the basis of such an isolated component, however thoroughly its functions may have been studied. The same applies to the phenomenon of heredity as well and, in general, to any property of living material.[63]

In June 1964, when Nuzhdin's name was put forward for membership in the Academy of Sciences, his supporters described his achievements in the following terms:

Nuzhdin had paid much attention to the problems of the struggle with anti-Michurinist distortions of biology, constantly criticizing various idealistic trends in the study of heredity and variation. His general philosophical works, in connection with the further development of the materialistic teaching of Michurin . . . are widely known.[64]

Nuzhdin's views deserve mention if only to illustrate that, behind the seemingly innocuous questions being posed, there lay deep ideological and political differences which loomed large in the cold war of the 1950s.

One must not overemphasize the political aspect of the meeting, however. The Soviet scientists did not form a homogeneous group of Lysenko supporters. Most of them supported the Oparin position without specific mention of Lysenko.[65] Olga Lepeshinskaia, by this time at odds with Oparin over his claim that the original forms of life could not exist in the present time, took part in the discussions during the conference. The doctrine that cells only arise from preexisting cells was disclaimed by her on the basis of her own laboratory findings, in which "non-cellular forms of life" had been transformed into bacterial cells.[66]

Oparin defined the distinction between his views and those of the geneticists very succinctly. "Is life only inherent in the individual molecule of protein, nucleic acid or nucleoprotein, and is the rest of the protoplasm merely a lifeless medium?" he asked. "Or is life inherent in a multimolecular system in which proteins and nucleic acids have an extremely important role, though it is that of a part, not that of a whole?" In Oparin's view the functionally efficient and highly structured molecules that one finds in living systems could never arise all at once by a spontaneous generation, but only through a long evolutionary process. In a rather delightful analogy, he likened his opponents' view to that of Empedocles, "who held that first there developed arms, legs, and eyes and ears and that later, owing to their combination, the organism developed."[67]

Oparin and Muller continued to exchange barbs well into the 1960s, and even today the issue remains unsolved. In 1961, Oparin once again attacked the idea that nucleic acid was somehow the first living molecule which arose by a fortuitous chemical combination:

The more concrete the biochemical studies of the self-reproduction of living beings, the more obvious it becomes that the process is not just bound up with this or that particular substance or a single molecule of it, but is determined by the whole system or organization of the living body which ... is flowing in nature and is in no way to be compared with a stamping machine with an unchanging matrix.[68]

In 1966, two years after the downfall of Nikita Khrushchev and the subsequent decline of Lysenko, Muller once more attacked Oparin's view and drew attention to his support of Lysenko:

It is a curious anachronism, however, that even today some of the most eminent biochemists and biologists, doing very valuable work in their respective fields, still adhere to this view [that protoplasm is primary] and its corollary concerning life's origin. Unfortunately, it became much publicized and elaborated, beginning in the 1930's, by the Lysenkoist Oparin in his book, *The Origin of Life* (1938 *et seq.*), as part of the attempt to down-rate the significance of genetics.[69]

The historical association of the origin of life question with the Lysenko-Mendel debate has led many modern biochemists to see the major question as: "Which had primacy in the origin of life, nucleic acids or proteins?"[70] This was not the basic issue, however. The fundamental question was whether life arose in one decisive step, as Muller postulated, or through a long and gradual process, as Oparin maintained.

Despite Oparin's association with the denigrated Lysenko and his hostility to much of modern genetics, Oparin's views on the origin of life have persisted. This can be attributed to the successful experimental testing of his four-stage process:

1. Appearance of hydrocarbons and cyanides and their immediate derivatives in cosmic space and during the formation of the earth and the subsequent development of its crust, atmosphere, and hydrosphere.

2. Conversion on the earth's surface of the initial carbon-containing compounds into more and more complex organic substances—monomers and polymers—appearance of the so-called "primordial soup."

3. The self-formation, in this soup, of polymolecular open systems capable of mutually interacting with the environment and capable of growth and multiplication on the basis of this interaction—appearance of probionts.

4. The further evolution of "probionts," the development of more perfect metabolism, more perfect molecular and super molecular structures accomplished through the basis of pre-

biological selection—the appearance of the primordial organisms.[71]

Indeed, the American biochemist, Sidney Fox, regarded "the problem in principle as more solved than unsolved."[72]

Today, the majority of biologists and biochemists seem committed to the evolutionary viewpoint of Oparin. Such commitment does not imply, of course, that there has been an encroachment of dialectical materialism into European and North American science. Rather, the evolutionary viewpoint has enabled the question of life's beginnings to be treated, for the first time, as a legitimate scientific research problem. On more theoretical grounds, the evolutionary approach has also received support from the publication of Harold Blum's *Time's Arrow and Evolution*. Curiously, however, although it had appeared in 1951, Blum's work was hardly mentioned during the 1957 International Symposium.

Blum, like Oparin, saw the origin of life as a transition between the nonliving and living, in which the spontaneous production of complex molecules, such as polypeptides, was thermodynamically "beyond all probability." He argued, also, that, were a group of scientists of different disciplines allowed to travel back in time to witness the beginning of life, there would be no universal agreement as to the exact point at which life began. On thermodynamic grounds, the origin of any one of the major properties of life, he argued, "is difficult enough to conceive, let alone the simultaneous origin of all." He concluded his chapter on the origin of life by suggesting that "we abandon the idea of a definite moment of origin and assume that a series of events represents the beginning of life rather than one definite point of this series."[73] With the publication of this work, Oparin's dialectical interpretation of the beginning of life became justified on the basis of predictions stemming from the second law of thermodynamics.

With the broad acceptance of Oparin's scheme, the issue of spontaneous generation should be resolved finally. Life did not arise by a spontaneous generation. That is to say, that a functional living entity, whether that be a mouse, maggot, bacterium, virus, or "living molecule," did not make an all-at-once appearance from material with no lifelike qualities. Life emerged slowly as part of a long developmental process, all stages of which were highly probable at the time they occurred. As such, it becomes meaningless to draw a line through these stages and to call stages below the line nonliving and those above living. It, therefore, also becomes meaningless to speak of a spontaneous generation of life, either today or in the past. Oparin's scheme allows for the evolutionary process to be a continuous one, but the very existence of life itself renders it impossible for the early stages to be repeated—all probionts would be destroyed by the bionts. Furthermore, argued Oparin:

If we can form a clear picture in our minds of the whole grandiose nature of this evolution, we must now regard as ludicrously naive the hopeless attempts which have been made to reproduce the spontaneous genesis of life in decaying decoctions and infusions of organic substances. The path followed by nature from the original systems of protobionts to the most primitive bacteria and algae was not in the least shorter or simpler than the path from the amoeba to man.[74]

There are those, however, who still believe in the sudden appearance of a *first,* functionally organized, living molecule, and who thus believe that life arose in the past by a spontaneous generation. Such an assumption is not only untestable but involves some dubious arguments. Its proponents stress that, although the chance combination of molecules which would produce a living entity was extremely improbable, it was not impossible. Thus, given the vast eons of time available before life appeared on the planet, the correct combination of molecules arising "by chance" becomes inevitable. In the words of George Wald: "One only has to wait: time itself performs the miracles."[75] The problem is, however, that as long as the presence or absence of life is known only for our planet, as long as the number of "trial" combinations of molecules is unknown, and as long as the time period remains unspecified, no meaningful statement as to the probability of life's arising can be made. That "time itself performs the miracles" is a truism and tells us nothing. An assessment of the probability that a complex entity such as DNA arose all at once demands that the time interval be specified. The possibility that *any* event can occur will approach one—that is, absolute certainty—as the length of time increases and, at infinite time, the possibility actually becomes one. Indeed, at infinite time, it becomes mathematically certain that a Shakespearean play will be composed by the fortuitous meeting of the letters of the alphabet. Similarly, if the time interval is lengthened so as to become "indefinite," then the possibility of any event's occurring will approach one asymptotically. It is possible by using such arguments—what Peter Mora has termed "the infinite escape clause"—to prove anything. Indeed, one can claim this probability-argument to be, with complete justification, the twentieth-century equivalent of Divine Creation:

These escape clauses postulate an almost infinite amount of time and an almost infinite amount of materials, so that even the most unlikely event could have happened. This is to invoke probability and statistical considerations when such considerations are meaningless. When for practical purposes the condition of infinite time and matter has to be invoked, the concept of probability is annulled. By such logic we can prove anything.[76]

The conflict between Oparin and Muller, between emergence and spontaneous generation, now rests on the relative assessment of two conflicting hypotheses. What is the more likely: that a complex living entity, having all the essential properties of life, arose suddenly, or that there has been a series of steps of increasing complexity, each step having a level of probability such that the occurrence of each would have been likely during the available time span? Although Oparin himself seems to have held a curiously anachronistic view of "chance," opposing spontaneous generation from much the same standpoint as eighteenth-century opponents, most modern biochemists and biologists have concluded that emergence is more probable than spontaneous generation. That a complex "living molecule" could have arisen all at once seems to be so improbable, having a probability so close to zero, that it appears virtually impossible. Time itself will perform that miracle, but adherence to such an hypothesis seems unnecessary when faced with a more-likely second hypothesis: that life emerged slowly. That possible stages in this gradual process have been reduplicated in the laboratory now provides additional armament for the Oparin hypothesis.

Unfortunately, many supporters of the evolutionary viewpoint still speak of "the spontaneous generation of life" and "the first organism," thereby confusing the whole issue. One of the most perplexing papers in this regard was George Wald's. While admitting that Oparin had provided the foundation for modern approaches to the problem and that "the origin of a living organism is undoubtedly a stepwise phenomenon, each step with its own probability,"[77] he nevertheless insisted that, "as natural scientists learn more about nature they are returning to a hypothesis their predecessors gave up almost a century ago: spontaneous generation." Although, clearly, we are dealing here with a matter of semantics, it only confuses the situation to state that the only alternative to a supernatural creation is belief in spontaneous generation and that today, "we now have to face a somewhat different problem: how organisms may have arisen spontaneously under different conditions in some former period, granted that they do so no longer."[78] Whether he admits to it or not, Wald is here using the language of the late nineteenth century, with all its ambiguities. There is indeed another alternative to that of creation or spontaneous generation, the alternative best expressed by the word "biopoesis"—the whole evolutionary process leading from inorganic beginnings to the emergence of life.

Unfortunately, the dialectical viewpoint, which fully supports such an approach, is itself subject to misunderstanding regarding spontaneous generation. The Soviet scientist A. S. Konikova criticized the use of the term "biopoesis" at the 1957 conference on the grounds that it failed to indicate a "qualitative break in nature." "The scheme provides a correct and consistent materialist picture of the form of development of nature but does not disclose the main point of these stages of development: it does not reflect the transition from chemistry to biology."[79] Indeed, recent work on self-assembly of macromolecules has illustrated that "major evolutionary changes may have yielded sudden leaps forward such as could not have been forecast from knowledge of the predecessors."[80] It makes little sense, and is confusing historically, to equate such leaps with the concept of spontaneous generation. To do so would imply that before such an event no lifelike processes were in operation and that after the leap all such processes had appeared.

Oparin himself could be misinterpreted in this way when he writes that life "is not separated from the rest of the world by an unbridgeable gap, but arises in the process of the development of matter, *at a definite stage* [emphasis added] of this development as a new, formerly absent quality."[81] However, in using such terminology he is referring to the stage at which new biological laws come into play and not to a sudden passage from a nonliving to a living state. Unfortunately, however, Oparin has continued to use the term "origin of life," rather than the less confusing "emergence of life," in all his writings, in an attempt to denote that there is a point in the evolutionary process at which qualitatively new biological laws come into operation. His continued use of this terminology has had the unfortunate effect of clouding over the fundamental difference between those who believe in the spontaneous generation of a first living molecular state and those who accept the evolutionary approach. Recently, Oparin has become more circumspect in the term's use. In 1968, he dropped the word *Origin* from the title of one of his books and used instead *Genesis and Evolutionary Development*,[82] and, in a more recent article, he seems to favor *appearance* and *emergence* over *origin*.[83]

Some recent papers have suggested that a recurring biopoesis may be a very distinct possibility; Keosian himself coined the term *neobiogenesis* for such events.[84] Unfortunately, Adolph Smith and Dean Kenyon have referred to such a possibility as "the contemporary spontaneous generation of life."[85] Once again we are dealing with an unfortunate usage, for they clearly accept the evolutionary viewpoint: they describe life's appearance as a "spontaneous generation spread over millions of years." Spontaneous generation, of course, means a sudden transition from a completely nonliving state to a fully-living state, which by definition cannot be spread over a long period of time.

However, Smith and Kenyon also make the quite fascinating claim that "both mycoplasma and viruses may be originating *de novo* within cells and tissue fluids of host organisms," and that "such newly formed units of life may play important roles in some disease processes." They hypothesize that "within the body fluids of host organisms, rapid back-and-forth movement is occurring between the states of organization of living matter," a suggestion that has some uncomfortably close relationships to previous claims in support of heterogenesis. They also suggest that the common notion of "infectious diseases," based as it is on the belief that the infectious agent comes from similar parents, needs reexamination. "In summary," they argue, "we are proposing that a new dimension be considered in biology, the back-and-forth movement between what is commonly considered life and death." These remarks cannot be said to represent a reappearance of the belief in spontaneous generation. Despite their use of the words *life* and *death,* it clearly makes little sense to label one of their states of organization "dead," and another "alive." It seems to provide one more example of unnecessary confusion brought about by adherence to outdated terminology.

Most, but not all, modern biologists and biochemists have abandoned any belief in spontaneous generation. This has come about through general acceptance of the evolutionary viewpoint as expressed by Aleksandr Oparin. Contrary to popular belief, therefore, the present state of the controversy has not resulted from disproof by infallible experimental evidence. The issue is one which has been abandoned many times before, only to reappear at a later date under a different guise. Whether the final chapter in the history of the spontaneous generation controversy has now been written it is impossible to say.

THE LIFE CYCLE OF
THE FLUKE

T he flukes, or digenetic trematodes, are found in the gut, liver, and other organs of vertebrates—especially fish, birds, and mammals. When mature, their eggs pass out of the vertebrate host, usually by way of the feces, to hatch into a small free-swimming larval stage called a *miracidium*. The miracidium then bores its way into the tissues of a suitable snail, the first intermediate host, in which it develops further into a succession of larval stages called *sporocysts, daughter sporocysts, rediae,* and *cercariae* respectively. The sporcysts are

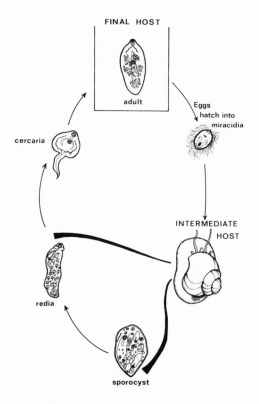

Life-cycle diagram of digenetic trematode or fluke.

saclike bodies which bud off into their lumen very many daughter sporocysts. When released from the sporocyst, each one of these in turn buds off internally many rediae, which in their turn bud off multitudes of cercariae. Thus, from one miracidium literally tens of thousands of cercariae are produced. The tailed cercariae migrate out of the rediae and bore their way out of the snail to live a short time as a free-swimming larval stage. Up until this stage, the life cycles of all trematodes are very similar, but thereafter they vary a good deal.

In the majority of trematodes the cercariae bore their way into the tissues of a second intermediate host, very often a fish, where they encyst as *metacercariae*. When eaten by the final host—a fish-eating bird, for example—the cyst breaks down and the immature fluke makes its way to the location within the host where it eventually matures to once again produce eggs.

In other cases, as in the famous sheep-liver fluke, the cercariae encyst on vegetation and mature after this vegetation is eaten by a sheep. In yet another case, there is no second intermediate host; the cercariae simply bore their way back into the final host where they mature.

Before the work of Japetus Steenstrup in 1842 it was not realized that the stages occurring within the snail were related to the fluke adult in the vertebrate body. Thus, it was widely assumed that both adult flukes and the snail stages were spontaneously generated. Since opponents of spontaneous generation assumed, quite naturally, that the fluke eggs hatched directly into the adult, they were unable to produce any arguments in their favor other than that of analogy.

APPENDIX TWO

THE LIFE CYCLE OF
THE TAPEWORM

The adult tapeworm is a ribbonlike being whose body is partitioned into numerous *proglottids*. The anterior end bears a scolex, by which the animal attaches itself to the gut wall, and the proglottids are budded off from the neck region immediately behind the scolex. Each proglottid bears a complete set of reproductive organs, those farthest from the scolex being the most mature. Tapeworms display two very different types of life cycles: cycles of those whose hosts are terrestrial, and the cycles of those whose hosts are aquatic.

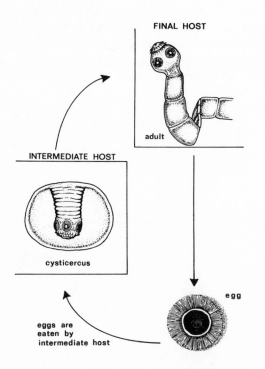

FINAL HOST

adult

INTERMEDIATE HOST

cysticercus

egg

eggs are
eaten by
intermediate host

Life-cycle diagram of terrestrial tapeworm in which all hosts are terrestrial and there are no free-swimming larval stages.

190

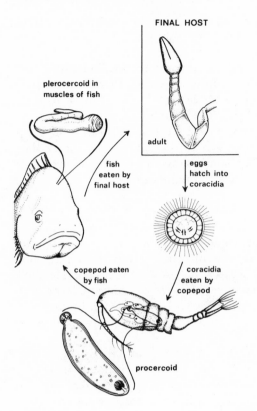

FINAL HOST

plerocercoid in
muscles of fish

adult

fish
eaten by
final host

eggs
hatch into
coracidia

copepod eaten
by fish

coracidia
eaten by
copepod

procercoid

Life-cycle diagram of aquatic tapeworm in which aquatic hosts are involved and a free-swimming larval stage is present.

TERRESTRIAL LIFE CYCLE.   Whole proglottids, containing many eggs, are shed with the feces of the final host. These thick-shelled eggs contain the six-hooked larval stages, which do not hatch until the eggs have been eaten by the first intermediate host, often another vertebrate. Then the larva hatches and bores its way into the tissues where it develops into a hollow, bladderlike cyst, the *cysticercus,* usually bearing a single inverted scolex. When this intermediate host is eaten by the final host, the scolex everts to attach to the gut wall and begins budding off proglottids.

Such a life cycle is illustrated by the human tapeworm, *Taenia solium,* whose cysticerci occur in the muscles of the pig. Historically, the species which drew the most attention was *Echinococcus,* whose larval stage, the so-called *hydatid cyst,* can occur in man. The adult *Echinococcus* occurs in carnivores, such as dogs, foxes, and wolves, while the hydatid larval stage is commonly found in herbivores such as sheep and cattle, but may also be found in man. The hollow hydatid cyst, which may reach 10 inches in diameter, differs from the more typical cysticercus in that the wall of the cyst buds off internally many daughter cysts. As these enlarge they may also, in turn, bud off internally tertiary cysts. Along the inner surfaces of both daughter and tertiary cysts, minute tapeworm scoleces develop. When the host containing the hydatid is eaten by predators, each one of these scoleces can attach to the gut wall and produce an adult worm.

Since these bladderlike cysts, the cysticerci and hydatids, lie deep within the body with no obvious way to the exterior and since they lack reproductive organs, they soon came to be seen as the most obvious example of a spontaneously generated organism.

AQUATIC LIFE CYCLE. In a typical case, the eggs, emitted with the feces of the final host, hatch in the water to produce free-swimming *coracidium* larvae. These larvae, when eaten by a suitable invertebrate, such as a copepod, bore their way into the body cavity where they form a small solid larval form or *procercoid*. When the copepod is eaten by a fish, a second larval stage, or *plerocercoid*, develops from the liberated procercoid. After these infested fish are eaten by humans, mammals, or birds, each plerocercoid develops into an adult tapeworm.

This type of life cycle is found in the human tapeworm, *Dibothriocephalus latus*, which people contract by eating infested fish. It is thus restricted in its distribution to the Baltic, European lakes, and the Great Lakes of North America. The different intermittent ranges of the terrestrial and aquatic tapeworms long puzzled scientists and seemed to provide one more argument in favor of their spontaneous generation.

NOTES

ONE. INTRODUCTION

1. Throughout the text I have used the term *spontaneous generation*. In the seventeenth and eighteenth centuries, it was common to use the term *equivocal generation* as opposed to *univocal generation*. Although in this text spontaneous generation and equivocal generation are considered synonymous terms, the literal meaning of the latter implies generation of an uncertain nature. This usage is understandable within the context of seventeenth and eighteenth century thought, when the possibility of spontaneous generation seemed so remote. Organisms, therefore, which on the surface seemed to arise without parents, were seen to reproduce in an ambiguous and uncertain manner. By the nineteenth century the terms spontaneous generation and equivocal generation were used interchangeably.

2. T. S. Kuhn, "Science: The history of science," in D. L. Sills, ed., *International Encyclopaedia of the Social Sciences* 14 (1968): 75.

3. This fairly typical account is taken from E. O. Wilson et al., *Life on earth* (Stamford, Conn.: Sinauer Associates, 1973), p. 593.

4. This inference is stressed by Everett Mendelsohn in his "Philosophical biology vs experimental biology: Spontaneous generation in the 17th century," *Actes XIIe Congrès International d'Histoire des Sciences* 1 (1971): 201–29.

5. J. R. Baker, *Abraham Trembley of Geneva: Scientist and philosopher* (London: Edward Arnold, 1952), p. 166.

6. C. Burdach, *Die Physiologie als Erfahrungswissenschaft*, 2nd ed. (Leipzig, 1832), vol. 1, p. 8.

7. H. C. Bastian, *The nature and origin of living matter* (London: Fisher Unwin, 1905), p. 147.

8. N. R. Hanson, *Patterns of discovery* (Cambridge: Cambridge University Press, 1963), p. 72.

9. J. W. S. Pringle, "The evolution of living matter," *New Biology* 16 (1954): 66.

TWO. ABHORRENCE OF CHANCE

1. *Lexicon technicum or a universal English dictionary of arts and sciences,* 1st ed. (London, 1704). Article on equivocal generation. This dictionary is a well-known compendium of knowledge of the period.

2. R. Descartes, "Formation de l'animal," in *Oeuvres*, ed. Ch. Adam and P. Tannery, 12 vols. (Paris, 1897–1913), 11: 277.

3. Quote from Walter Pagel, *Paracelsus* (Basel: S. Karger, 1958), p. 91.

4. F. Redi, *Experiments on the generation of insects* (1688), English trans. M. Bigelow (Chicago, 1909), p. 23.

5. P. M. Rattansi, "The Helmontian-Galenist controversy in Restoration England," *Ambix* 12 (1964): 1–23.

6. P. M. Rattansi, "Paracelsus and the Puritan revolution," *Ambix* 11 (1963): 24–32.

7. This is discussed by C. Webster in "Water as the ultimate principle of nature: The background to Boyle's Sceptical Chymist," *Ambix* 13 (1966): 96–107.

8. J. B. van Helmont, *Oriatrike, or physick refined*, trans. J. Chandler (London, 1662), p. 110.

9. J. Roger, *Les Sciences de la vie dans la pensée française du XVIIIe siècle* (Paris: Armand Colin, 1963), p. 121.

10. Rattansi, "Helmontian-Galenist controversy," p. 21.

11. T. M. Brown, "The College of Physicians and the acceptance of iatromechanism in England, 1665–1695," *Bull. Hist. Med. 44* (1970): 12–30.

12. J. de Savants, 1693, quoted in Roger *Les Sciences de la vie*, p. 207.

13. G. Garden, "A discourse concerning the modern theory of generation," *Phil. Trans. Roy. Soc. 16* (1691): 476.

14. "An account of the dissection of a bitch, whose *Cornua Uteri* being filled with the bones and flesh of a former conception, had after a second conception the ova affix't to several parts of the abdomen," *Phil. Trans. Roy. Soc. 13* (1683): 186.

15. T. H. Huxley, "Biogenesis and abiogenesis" (1870), in *Discourses biological and geological* (New York, 1894), p. 240.

16. The quote is from a chapter heading in Roger, *Les Sciences de la vie*, p. 163.

17. A. Vallisneri, *Opere fisciso-mediche* (Venice, 1733), vol. 1, p. 132, quoted in Roger, *Les Sciences de la vie*, p. 209.

18. Roger, *Les Sciences de la vie*, p. 213.

19. K. Digby, *Two treatises, in the one of which the nature of bodies; in the other, the nature of man's soule is looked into: in way of discovery of the immortality of reasonable soules* (Paris, 1644), p. 215.

20. This point is made in Roger, *Les Sciences de la vie*, and in P. Bowler, "Preformation and preexistence in the seventeenth century: A brief analysis," *J. Hist. Biol.* 4 (1971): 221–44.

21. "Gassendi on generation," in H. B. Adelmann, *Marcello Malpighi and the evolution of embryology* (Ithaca N.Y.: Cornell University Press, 1966), vol. 2, pp. 798–816.

22. E. Gasking, *Investigations into generation, 1651–1828* (Baltimore: Johns Hopkins Press, 1967); E. Mendelsohn, "Philosophical biology vs experimental biology: Spontaneous generation in the seventeenth century," *Actes XIIe Congrès International d'Histoire des Sciences* 1 (1971): 201–29; G. Keynes, *The life of William Harvey* (Oxford: Clarendon Press, 1966), p. 352.

23. W. Harvey, *Anatomical exercises on the generation of animals,* in *The works of William Harvey*, trans. R. Willis (London: Sydenham Society, 1847), p. 462. These aspects of Harvey's work are discussed further in Gasking, *Investigations into generation.*

24. Harvey, *On animal generation*, p. 463.

25. Ibid., p. 340.

26. T. Kerckring, "An account of what hath been of late observed by Dr. Kerckringius concerning eggs to be found in all sorts of females," *Phil. Trans. Roy. Soc.* 7 (1672): 4018–26.

27. Ibid., p. 4024.

28. R. de Graaf, *De mulierum organis generationi inservientibus tractatus novus, demonstrans tam homines et animalia caetera omnia, quae vivipara dicunter, haud minus quam ovipara, ab ovo originem ducere* (Leiden, 1672).

29. J. de Savants, 6 March 1679; quotation from Roger, *Les Sciences de la vie*, p. 267.

30. Redi, *Experiments on generation*, p. 26.

31. Ibid., p. 33.

32. Ibid., p. 90.

33. Ibid., pp. 92, 116.

34. J. Swammerdam, *Histoire générale des insectes* (Utrecht, 1682), p. 159.

35. Ibid., p. 47.

36. Roger, *Les Sciences de la vie*, p. 241.

37. J. Ray, *The wisdom of God manifested in the works of creation*, 1st ed. (London, 1691), p. 28.

38. Ibid., 2nd ed. (London, 1692), pp. 73–75.

39. N. Le P. Malebranche, *De la recherche de la vérité* (Paris, 1673), bk. 1, chap. 6.

40. Bowler, "Preformation and preexistence." Metamorphosis was the belief that all parts of the embryo appear simultaneously *after* conception.

41. "Extrait d'une lettre de M. Huygens de l'Académie Royale des Sciences," *J. de Savants*, 15 August 1678.

42. "Extrait d'une lettre de M. Nicolas Hartsoeker," *J. de Savants*, 29 August 1678.

43. A. van Leeuwenhoek, "The observation of Mr. Anthony Leeuwenhoek, on animalcules engendered in the semen," *Phil. Trans. Roy. Soc.* (1679), letter 22.

44. A. van Leeuwenhoek, *The collected letters* (Amsterdam: Swets and Zeitlinger, 1941), letter to Grew, 18 March 1678, vol. 2, p. 327.

45. Ibid., letter to Hooke, 5 April 1680, vol. 3, p. 203; discussed further in C. Castellani, "Spermatozoan biology from Leeuwenhoek to Spallanzani," *J. Hist. Biol.* 6 (1973): 37–68.

46. Garden, "Modern theory of generation."

47. J. Ray, *The wisdom of God*, p. 221.

48. R. Owen, *Lectures on the comparative anatomy and physiology of the invertebrate animals* (London, 1843), p. 54.

49. N. Andry de Boisregard, *De la génération des vers dans le corps de l'homme*, 1st ed. (Paris, 1700); an extended third edition appeared in Paris in 1741. All quotations are from that 3rd edition; p. 11.

50. Ibid., p. 844.

51. Ibid., p. 31.

52. Ibid., p. 714, Hartsoeker's letter to Andry de Boisregard.

53. A. Vallisneri, *New observations and experiments upon the eggs of worms found in humane bodies* (London, 1713), partial trans. in D. LeClerc, *A natural and medicinal history of worms* (London, 1715), trans. from Latin J. Browne (London, 1721), p. 351.

54. Ibid., p. 353.

55. Ibid., p. 347.

56. For a more detailed discussion of these issues, see Gasking, *Investigations into generation*, and A. Lovejoy, *The great chain of being* (Cambridge, Mass.: Harvard University Press, 1953).

57. P.-L. de Maupertuis, *Vénus physique* (1745); English trans. S. B. Boas (New York: Johnson Reprint, 1966), p. 56.

58. G.-L. de Buffon, *Histoire naturelle, générale et particulière*, 1st ed. (Paris, 1749), vol. 2, *Histoire générale des animaux*, pp. 24–32.

59. G.-L. de Buffon, *Oeuvres complètes*, ed. M. Lamouroux (Paris, 1828), vol. 12, *Histoire générale des animaux*, p. 373. This section dealing with the spontaneous generation of parasitic worms was included in all editions of *Histoire naturelle* which appeared after 1802.

60. Buffon, *Histoire naturelle*, 1st ed., vol. 2, p. 320.

61. J. T. Needham, "A summary of some late observations upon the generation, composition, and decomposition of animal and vegetable substances," *Phil. Trans. Roy. Soc.* 45 (1748): 615–66.

62. Ibid., par. 6.

63. Ibid., par. 10.

64. Ibid., par. 29.

65. Ibid., par. 31.

66. Ibid., par. 21.

67. Ibid., par. 4.

68. L. Spallanzani, *Nouvelles recherches sur les découvertes microscopiques et la génération des corps organisés* (London and Paris, 1769), p. 134.

69. Ibid.

70. L. Spallanzani, *Tracts on the natural history of animals and vegetables*, English trans. J. G. Dalyell (Edinburgh, 1803), p. 1.

71. Ibid., p. 9.

72. Ibid., p. 13.

73. Ibid., p. 13.

74. Ibid., p. 15.

75. Ibid., p. 16.

76. Ibid., p. 27.

77. Spallanzani, *Nouvelles recherches*, pp. 1–2.

78. C. Bonnet, *Contemplation de la nature*, in *Oeuvres d'histoire naturelle et de philosophie* (Neuchâtel, 1779), vol. 9, chap. 1, p. 2.
79. C. Bonnet, *Considérations sur les corps organisés* (Amsterdam, 1762), p. 115.
80. C. Bonnet, "Dissertation sur le Taenia," in *Oeuvres*, vol. 3, p. 96.
81. Quotation from A. Vartanian, *Diderot and Descartes* (Princeton, N.J.: Princeton University Press, 1953), p. 51.
82. Ibid., p. 259.
83. D. Diderot, *Philosophic thoughts* (1746), in L. G. Crocker, ed., *Diderot's selected writings* (New York: Macmillan, 1966), p. xix.
84. D. Diderot, *Le Rêve de d'Alembert* (Paris: Librairie Marcel Didier, 1951), pp. 14–15.
85. Ibid., p. 25.
86. Ibid., p. 55.
87. Ibid., p. 59.
88. Gasking, *Investigations into generation*, chap. 8.

THREE. THE TRIUMPHANT AGE OF SPONTANEOUS GENERATION

1. Examples of such awareness are seen in, E. Mendelsohn, "The biological sciences in the 19th century: Some problems and sources," *Hist. of Science* (1964): 39–59; L. Pearce Williams, *The origins of field theory* (New York: Random House, 1966); and B. Gower, "Speculation in physics: The history and practice of *Naturphilosophie*," *Stud. Hist. Phil. Sci.* 3 (1973): 301–56.
2. The following discussion of *Naturphilosophie* is taken from A. Höffding, *A history of modern philosophy*, vol. 2 (London, 1908); and A. Lovejoy, *The reason, the understanding, and time* (Baltimore: Johns Hopkins Press, 1961). The latter appears to me to be the only intellible account of *Naturphilosophie* available in the English language.
3. T. Carlyle, "The state of German literature," *Edinburgh Review* (1827): 348. Quoted in Lovejoy, *Reason, understanding, and time*, p. 16.
4. F. Schelling, quoted in Lovejoy, *Reason, understanding, and time*, p. 56.
5. F. Schelling, *Von der Weltseele*, in *Sämmtliche Werke*, part 1, vol. 2, p. 348. Quoted in Lovejoy, *Reason, understanding, and time*, p. 115.
6. A. Lovejoy, "Herder: Progressionism without transformism," in B. Glass, O. Temkin and W. L. Straus, Jr., eds., *Forerunners of Darwin* (Baltimore: Johns Hopkins Press, 1959), chap. 8.
7. C. G. Carus, "Von den Naturreichen, ihren Leben und ihrer Verwandtschaft," *Zeit für Natur u Heilkunde* 1 (1819): 1–72. English trans. in Richard Taylor, *Scientific Memoirs* (London, 1837), Vol. 1, p. 224.
8. Ibid., p. 225.
9. G. Treviranus, *Biologie oder Philosophie der lebenden Natur* (Göttingen, 1805), vol. 3.
10. L. Oken, *Lehrbuch der Naturphilosophie* (Jena, 1809), pars. 946 and 865 respectively.
11. M. E. Bloch, *Abhandlung von der Erzeugung der Eingeweidewürmer und den Mitteln wider dieselben* (Berlin, 1782); French trans., *Traité de la génération des vers des intestines et des vermifuges* (Strasbourg, 1788), p. 84.
12. Ibid., p. 83.
13. J. G. Bremser, *Über lebende Würmer im lebenden Menschen* (Vienna, 1819), p. 30. A poor inexact French translation appeared in Paris in 1837, *Traité zoologique et physiologique sur les vers intestinaux de l'homme*. Bremser (1767–1827) graduated from Jena and then specialized in the study of helminthology at the Natural History Museum of Vienna. Between 1815 and 1825 he worked in Paris stressing the medical and more practical aspects of parasitology.
14. Ibid., pp. 108, 109.
15. Ibid., p. 38.
16. Ibid., p. 49.

17. Ibid., p. 51.
18. Ibid., p. 53.
19. Ibid., p. 64.
20. Ibid., p. 61.
21. A. Dugès, *Traité de physiologie comparée de l'homme et des animaux* (Paris, 1838). An extensive discussion in support of spontaneous generation appears in volume 3, on page 197.
22. This brief discussion of Cuvier is drawn from Coleman, *Georges Cuvier, zoologist* (Cambridge, Mass.: Harvard University Press, 1964).
23. G. Cuvier, *Rapport historique sur les progrès des sciences depuis 1789 et sur leur état actuel* (Paris, 1810), p. 193.
24. G. Cuvier, "Sur un nouveau rapprochement à établir entre les classes qui composent le règne animal," *Ann. du Muséum* 19 (1812): 76. Quoted in Coleman, *Georges Cuvier*, p. 92.
25. Cuvier contemptuously cast aside the problem of parasitic worms in his discussion of the class Intestina of the division Radiata. In *The Animal Kingdom* (trans. E. Griffith [London, 1834], p. 457), he wrote:
   The difficulty of conceiving how they get into such places, joined to the observation that they never appear outside living bodies, has caused some naturalists to imagine that they engender spontaneously. It is now, however, quite certain, not only that the majority manifestly produce eggs, or living young, but that many of them have separate sexes and couple like ordinary animals. We may then believe that they propagate by germs, sufficiently small to be transmitted through the narrowest passages, or that frequently the animals in which they live bring the germs into the world with them.
26. An analysis of the writings of Jean-Baptiste Lamarck occurs in R. W. Burkhardt, Jr., "The evolutionary thought of Jean-Baptiste Lamarck" (Doctoral thesis, Harvard University, 1972). Also by the same author, "The inspiration of Lamarck's belief in evolution," *J. Hist. Biol.* 5 (1972): 413–38.
27. J.-B. Lamarck, *Système des animaux sans vertèbres* (Paris, 1801), p. 409.
28. J.-B. Lamarck, *Flora Françoise* (Paris, 1778), vol. 1, p. 2.
29. J.-B. Lamarck, *Mémoires de physique et d'histoire naturelle* (Paris, 1797), p. 272.
30. J.-B. Lamarck, *Philosophie zoologique* (Paris, 1809), vol. 2, p. 96.
31. Ibid., vol. 1, p. 384. Despite a belief in spontaneous generation, he wrote, "There is then, between brute or inorganic bodies and living bodies, an enormous difference, a considerable hiatus, in a word, a separation such that no inorganic body whatsoever can be approached by the simplest of living bodies."
32. Ibid., p. 210.
33. Ibid., vol. 2, p. 86. In the same chapter he asserts that animal life arises from a gelatinous mass, plant life from a mucilaginous mass.
34. Ibid., vol. 1, p. 373.
35. Lamarck, *Système des animaux sans vertèbres*, p. 331.
36. For details see J. Farley, "The initial reactions of French biologists to Darwin's *Origin of Species*," *J. Hist. Biol.* 7 (1974): 275–300.
37. *Oeuvres complètes de Cabanis* (Paris, 1824), Mémoire 10, vol. 4., p. 253n.
38. C. H. Duzer, *Contributions of the idéologues to French revolutionary thought* (Baltimore: Johns Hopkins Press, 1935). Also A. Dansette, *Religious history of modern France*, trans. J. Dingle, 2 vols. (Freiburg, 1961).
39. J. M. Barry, "On the origin of intestinal worms, particularly *Ascaris vermicularis*," *Trans. Assoc. of Fell and Licent. Kings and Queens Coll. of Physicians in Ireland* 2 (1818): 383–96.
40. W. Rhind, "An examination of the opinions of Bremser and others on the equivocal production of animals," *Edinburgh J. Nat. and Geol. Sci.* 2 (1830): 391–97.
41. J. Priestley, "Observations on different kinds of air," *Phil. Trans. Roy. Soc.* 62 (1772): 193.
42. J. Priestley, "On the spontaneous emission of dephlogisticated air from water containing a vegetating green matter," in *Experiments and observations on different kinds of*

air, and other branches of natural philosophy (Birmingham, 1790; New York: Kraus Reprint, 1970), p. 288.

43. J. Priestley, "On the putrefaction of air by plants and the influence of light on that process," ibid., pp. 293–305. Also reported in a letter to Giovanni Fabroni, September 1779, in R. Schofield, *A Scientific biography of Joseph Priestley, 1773-1804* (Cambridge, Mass.: M.I.T. Press, 1966. He accounted for his earlier mistake by claiming that seeds were either present in the water or passed through the cork. Later he prevented the entry of air-borne seeds by covering with mercury.

44. Priestley, "On spontaneous emission," p. 290.

45. J. Ingenhousz, *Experiments upon vegetables* (London, 1779), reprinted in *Chronica Botanica* 11 (1947): 325.

46. Ibid., p. 368. Both Carus, "Von den Naturreichen," and Treviranus, *Biologie oder Philosophie*, made specific reference to green matter in their discussions of spontaneous generation.

47. Ingenhousz, *Experiments upon vegetables*, p. 381.

48. J. Priestley, "Observations and experiments relating to equivocal or spontaneous generation," *Trans. Amer. Phil. Soc.* 6 (1809): 121.

49. E. Darwin, *The temple of nature or, the origin of society: A poem* (New York, 1804), line 231. He referred to Priestley's green matter in "Additional notes," pp. 142 and 147.

50. Priestley, "Observations and experiments," p. 123.

51. Ibid., p. 124.

52. Ibid., p. 125. Priestley's opposition to spontaneous generation is disccussed in H. J. Abrahams, "Priestley answers the proponents of abiogenesis," *Ambix* 12 (1964): 44–71.

53. C. F. Burdach, *Die Physiologie als Erfahrungswissenschaft* 2nd ed. (Leipzig, 1832), vol. 1, p. 8.

54. Canabis, *Oeuvres complètes*, p. 242.

FOUR. DOUBT AND UNCERTAINTY

1. J. Farley, "The spontaneous generation controversy (1700–1860): The origin of parasitic worms," *J. Hist. Biol.* 5 (1971): 113. My account of parasites here is based on this paper.

2. J. Müller, *Elements of physiology* (1835), English trans. W. Baly (London, 1838), p. 27.

3. Ibid., p. 16.

4. E. Mendelsohn, "Physical models and physiological concepts: Explanation in nineteenth century biology," *Brit. J. Hist. Sci.* 2 (1965): 203–19.

5. The statement is Ludwig's. An excellent account of the impact of this school is given in P. F. Cranefield, "The organic physics of 1847 and the biophysics of today," *J. Hist. Med.* 12 (1957): 407–23. Ludwig's remark is quoted on page 407 of the account.

6. H. von Helmholtz, "Aim and progress of physical science," in *Popular lectures on scientific subjects*, English trans. E. Atkinson (New York, 1873), p. 384.

7. T. Schwann, *Isis*, von Oken (1837), vol. 5, column 524.

8. T. Schwann, "Vorläufige Mittheilung betreffend Versuche über die Weingährung und Fäulniss," *Ann. d. Physik. u. Chemie* 41 (1837): 184–93. An account of these experiments is given in W. Bulloch, *The history of bacteriology* (London: Oxford University Press, 1938).

9. H. von Helmholtz, "Über das Wesen der Fäulniss und Gährung," *Arch. f. Anat. Physiol. u. Wissen. Med* (1843): 453–62.

10. For excellent histories of fermentation, see A. Harden, *Alcoholic-fermentation*, 4th ed. (London: Longman's Green, 1932); and J. S. Fruton, *Molecules and life* (New York: John Wiley, 1972), pp. 22–86.

11. A. Lavoisier, *Elements of chemistry* (1789), trans. R. Kerr (New York: Dover, 1965), p. 139.

12. J. von Liebig, *Chemistry and its application to agriculture and physiology* (1840), English trans. L. Playfair (Philadelphia, 1847), pp. 90 and 96.

13. C. Cagniard-Latour, "Mémoire sur la fermentation vineuse," *Ann. Chimie. et Phys.* 68 (1838): 221. Also, F. Kützing, "Microscopische Untersuchungen über die Hefe und Essigmutter nebst mehreren andern dazu gehörigen vegetabilischen Gebilden," *Zeit. f. prakt. Chemie* 11 (1837): 385–409.

14. Schwann, " Weingährung und Fäulniss," p. 192.

15. H. Schroeder and T. von Dusch, "Über Filtration der Luft in Beziehungen auf Fäulniss und Gährung," *Ann. de Chemie u. Pharm.* 89 (1854): 232–43.

16. H. Schroeder, "Über Filtration der Luft in Beziehung auf Fäulniss, Gährung und Krystallisation," Ibid., 109 (1859): 35–52; and 117 (1861): 273–95.

17. T. Schwann, *Microscopical researches into the accordance in the structure and growth of animals and plants*, English trans. H. Smith (London, 1847), p. 39.

18. Ibid., p. 190.

19. Ibid., p. 165.

20. K. Nägeli, "On the nuclei, formation, and growth of vegetable cells," Part II, in *Reports and papers on botany* (London: Ray Society, 1849), p. 134.

21. H. von Mohl, *Principles of the anatomy and physiology of the vegetable cell*, English trans. A. Henfrey (London, 1852) p. 57.

22. Ibid., p. 60.

23. "Cellular pathology: Its present position," *Brit. Med. J.*, 1861 Jan. 12, p. 44.

24. L. Büchner, *Kraft und Stoff*, 5th ed. (Frankfurt, 1858), p. 70. During this period, increasing knowledge of the geological record led to speculations on the origin of species. While most geologists tended to ignore the issue (*see* chapter five) or accepted the idea of supernatural creations, a few biologists believed new species arose by a series of spontaneous generations. They claimed support for this view by reference to Schwann's theory of exogeny. In 1843, for example, Stephan Endlicher and Franz Unger maintained that species may have arisen "from a slimy carbonaceous substratum," if, as they remarked, "a conclusion based on present experiences is admissible." See O. Temkin, "The idea of descent in post-romantic German biology: 1848–1858," in B. Glass et al., eds., *Forerunners of Darwin: 1745–1859* (Baltimore: Johns Hopkins Press, 1959), p. 331.

25. R. Remak, "Über extracellulare Entstehung thierischer Zellen und über Vermehrung derselben durch Theilung," *Archiv. f. Anat. Physiol. wissen. Med.* (1852): 49.

26. R. Virchow, "Cellular pathology" (1855), in *Disease, life, and man: Selected essays by R. Virchow* (Stanford, Cal.: Stanford University Press, 1958), p. 102. In an essay of 1893, "The place of pathology among the biological sciences," he remarked, "Inasmuch as he [Schwann] asserted that new cells originated from unformed matter, from 'cytoblastema,' the door was thrown open to the old doctrine of *generatio equivoca.*" Ibid., pp. 165–83.

27. R. Virchow, "Alter und neuer Vitalismus," *Archiv. f. Pathol. Anat. und Physiol.* 9 (1856): 23.

28. Virchow; "Cellular pathology," p. 98.

29. R. Virchow, *Die Cellularpathologie* (1858), English trans. F. Chance (New York, 1863), p. 55.

30. C. G. Ehrenberg, "Über das Entstehen des Organischen aus einfacher sichtbarer Materie, und über die organischen Molecüle und Atomen insbesondere, als Erfahrungsgegenstände," *Ann. d. Physik u. Chemie* 24 (1832): 1–48. English translation appeared in Richard Taylor, *Scientific Memoirs*, vol. 1 (London, 1837), pp. 555–83.

31. F. Dujardin, *Histoire naturelle des zoophytes* (Paris, 1841), p. 26.

32. D. Eschricht, "Enquiries, experimental and philosophical, concerning the origin of intestinal worms," *Edin. New. Phil. Journal* 31 (1841): 314–56.

33. M. Foucault, *The birth of the clinic*, English trans. A. M. Sheridan Smith (New York: Random House, 1973). Xavier Bichat remarked in the preface to *Anatomie général, appliquée à la physiologie et la médecine* (Paris, 1801), "Open up a few corpses: you will dissipate at once the darkness that observations alone could not dissipate."

34. R. T. H. Laennec, *A treatise on the diseases of the chest* (1819), English trans. J. Forbes, 1921 (New York: Hafner Publishing Co., 1962), preface.

35. Figures obtained from bibliography to T. S. Cobbold, *Entozoa: An introduction to the study of helminthology* (London, 1864).

36. R. Bright, "Observations on abdominal tumors and intumescence: Illustrated by some cases of acephalocyst hydatids," Guy's Hospital Report No. V (1840); reprinted in *Clinical memoirs on abdominal tumors and intumescence* (London, 1860), p. 11.

37. B. Phillips, "Cysts," in R. B. Todd, *The cyclopaedia of anatomy and physiology* (London, 1835–1839), vol. 1, p. 790.

38. A. Thomson, "Generation," in ibid., vol. 2, p. 431.

39. J. Goodsir, "On the anatomy and development of the cystic entozoa," (1844), in *Anatomical memoirs*, W. Turner, ed. (Edinburgh, 1868), vol. 2, p. 486.

40. G. Gros, "De la génération spontanée ou primitive en général et en particulier des helminthes," *Bull. Soc. Imp. Naturalists d. Moscou* 20 (1847): 517–40. A beautiful picture of this confusion was given by C. Burdach, *Die Physiologie als Erfahrungswissenschaft* (Leipzig, 1832), vol. I:

> Bojanus has discovered some yellow worms in the liver of snails [Distomes?], in which lived some cercariae.... Carus discovered, in the same organ, a worm (*Leucochloridium*) which was full of distome eggs. Then Siebold discovered that one worm which lived in birds (*Monostomum mutabile*) already contained in the egg another embryonic worm similar to those already seen by Bojanus. Thus the bird nursed a monostome in which was an egg containing a young monostome serving itself to house a distome!

41. R. Leuckart, *The parasites of man, and the diseases which proceed from them,* trans. W. E. Hoyle (Edinburgh 1886), p. 29.

42. C. T. von Siebold, *Über Band-und Blasenwürmer* (1854), English trans. T. H. Huxley, *Tape and cystic worms* (London: Sydenham Society, 1857), p. 12.

43. J. J. S. Steenstrup, *On the alternation of generations; or the propagation and development of animals through alternate generations,* trans. G. Busk (London: The Ray Society, 1845), p. 1.

44. Ibid., p. 3.

45. Ibid., p. 65.

46. C. T. von Siebold, "Bericht über die Leistungen im Gebiete der Helminthologie während des Jahres 1842," *Archiv. f. Naturgesch* 9 (1843): 300–35.

47. F. Dujardin, *Histoire naturelle des helminthes ou vers intestinaux* (Paris, 1845), p. 477.

48. C. T. von Siebold, "Parasiten," in R. Wagner, *Handwörterbuch der Physiologie* (Brunswick, 1844), p. 640.

49. Dujardin, *Histoire des helminthes,* p. 544.

50. F. Küchenmeister, *Die Todtenbestattungen der Bibel und die Feuerbestattung* (Stuttgart, 1893), pp. v–x.

51. F. Küchenmeister, *Über Cestoden im Allgemeinen und die des Menschen insbesondere* (Zittau, 1853), pp. 11–12.

52. Ibid., p. 19.

53. Ibid., p. 10.

54. P. J. van Beneden, *Les Vers cestoides ou acotyles, considérés sous le rapport de leur classification, de leur anatomie et de leur développement* (Brussels: Académie Royale de Belgique, 1850), p. 100.

55. These are tapeworms whose adult stages occur usually in fish-eating birds and mammals. Their larval stages, which occur in aquatic invertebrates and fishes, are solid in construction. The human tapeworm *Dibothriocephalus* belongs to this order. The typical degenerate looking bladderlike larvae belonging to the order Cyclophyllidea, are restricted to terrestrial hosts. They include the famous beef and pork tapeworm of the genus *Taenia*. In the case of *Cysticercus fasciolaris,* the bladder is small and the scolex is carried at the end of the long neck. Hence it looks less degenerate than the more typical cysticerci. (*See* appendix two.)

56. C. T. von Siebold, "Uber den Generationswechsel der Cestoden nebst einer Revision der Gattung Tetrarhynchus," *Zeit. f. wissen. Zool.* 2 (1850): 222.

57. Küchenmeister, "Einiges über den Übergang der Finnen in Taenien und über das Digitalin," *Gunsburg Zeit. Klin. Med.* 2 (1851): 295–99. Also, there is a short "Vorläufige Mittheilung" on page 240.

58. A delightful insight into the power and prestige of the German Professor is provided in R. B. Goldschmidt, *The golden age of zoology* (Seattle: University of Washington Press, 1956).

59. Küchenmeister, "Übergang der Finnen." In "Vorläufige Mittheilung," he identified the tapeworm as *Taenia crassiceps*, but in the major article as *T. serrata*. This error he explained was "understandable in one who is self-taught." In "Über die Umwandlung der Finnen und Bandwürmer," *Vierteljahrs f. prakt. Heilk.* (Prague) 1 (1852): 150, he changed his mind again to erect the new species *T. pisiformis*.

60. C. T. von Siebold, "Über die Verwandlung des *Cysticercus pisiformis* in *Taenia serrata*," *Zeit. f. wissen. Zool.* 4 (1853): 401.

61. Ibid., and "Über die Umwandlung der Blasenwürmer in Bandwürmer," Communique to the Breslau Society in 1852, French trans. in *Ann. Sci. Nat.* 17 (1852): 377–81.

62. von Siebold, *Tape and cystic worms*, chap. 4, p. 76.

63. R. Leuckart. "Lettre relative à de nouvelles expériences sur le développement des vers intestinaux." *Ann. Sci. Nat.* 3 (1855): 354.

64. P. J. van Beneden, "Mémoire sur les vers intestinaux," *Comptes-rendus Acad. Sci.* (Paris), supp, 1861.

65. Such claims were made in his 1861 Mémoire and in his "La Génération alternante et la digenèse," *Brux. Roy. Acad. Sci. Bull.* 20 (1853): 10–23. He claimed that the fundamental aspect of the alternation of generations was not simply a change in the "form and nature" of the organisms, as Steenstrup had claimed, but rather an alternation of reproductive methods—sexual and asexual. Not surprisingly Streenstrup responded to this belittlement, *Réclamation contre la génération alternante et la digenèse par Prof. P. J. van Beneden* (Copenhagen, 1854). A recent discussion of these events by E. Lagrange, "Le Centenaire d'une découverte: Le Cycle évolutif des Cestodes," *Ann. de Parasitol.* 27 (1952): 557–70, takes a rather less jaundiced approach to Beneden's claims than I do here.

66. E. Herbst, "Beobachtungen über *Trichina spiralis*," *Nachn. kgl. Ges. Wiss.* (Göttingen) 19 (1851): 260–64.

67. E. G. Reinhard, "Landmarks of Parasitology, II: Demonstration of the life-cycle and pathogenicity of the spiral threadworm," *Expt. Parasitol.* 7 (1958): 108–23.

68. Virchow, *Cellularpathologie*, p. 54.

69. L. Büchner, *Kraft und Stoff*, 1st ed. (Frankfurt, 1855), p. 72.

70. Ibid., 5th ed. (Frankfurt, 1858), foreword to 4th ed., p. xlix.

71. Ibid., 1st ed., p. 73.

72. J. Müller, *Elements of physiology*, English trans. W. Baly (London, 1838), p. 9.

73. J. H. Brooke, "Wöhler's urea, and its vital force?—A verdict from the chemists," *Ambix* 15 (1968): 104.

74. T. O. Lipman, "Vitalism and reductionism in Liebig's physiological thought," *Isis* 58 (1967): 176.

75. J. von Liebig, "Letter 23," in *Familiar letters on chemistry in its relation to physiology, dietetics, agriculture, commerce and political economy*, ed. J. Blyth, 4th ed. (London, 1859), p. 285.

76. Büchner, *Kraft und Stoff*, 5th ed., preface to 4th ed., p. xlix.

77. C. Vogt, *Natürliche Geschichte der Schöpfung des Weltalls* (Brunswick, 1851), pp. 135, 136. Vogt's views on spontaneous generation are discussed in F. G. Gregory, *Scientific materialism in 19th century Germany* (Doctoral thesis, Harvard University, 1973). My brief account is taken from this work, pages 314–20.

78. C. Vogt, *Bilden aus dem Thierleben* (Frankfurt, 1852), pp. 109–12.

79. C. Vogt, *Vorlesungen über den Menschen* (Giessen, 1863), vol. 2, p. 253.

80. R. Chambers, *Vestiges of the natural history of creation* (New York, 1845), p. 127.

81. R. Chambers, *Explanations: A sequel to "Vestiges of the natural history of creation"* (New York, 1846), pp. 19-20.

82. A letter from Weekes to Chambers, dated 2 September 1845, in which the experiment is described in detail, is included in ibid., pp. 134-39.

83. W. Paley, *Natural theology or evidences of the existence and attributes of the Deity* (London, 1802; reprint, Gregg International Publishers, 1970), p. 12.

FIVE. POPULARITY RESTORED

1. L. Büchner, *Force and Matter*, 8th ed. English trans. J. F. Collingwood (London, 1864), Introductory letter by Büchner, p. ix.

2. L. Büchner, *Kraft und Stoff*, 5th ed. (Frankfurt, 1858), foreword to 4th edition, p. xlix.

3. Ibid., 1st ed. (Frankfurt, 1855), p. 73.

4. M. Schultze, "Über Muskelkörperchen und das, was man eine Zelle zu nennen habe," English trans. in T. S. Hall, *A source book in animal biology* (Cambridge, Mass.: Harvard University Press, 1970), p. 451.

5. E. Haeckel, *Natürliche Schöpfungs-Geschichte* (Berlin, 1868); English trans. E. Lankester as *The history of creation or the development of the earth and its inhabitants by the action of natural causes* (New York, 1876), p. 345.

6. J. Cleland, "On cell theories," *Quart. J. Micro. Sci.* 13 (1873): 258.

7. J. Drysdale, *The protoplasmic theory of life* (London, 1874), p. 4.

8. T. H. Huxley, "On the physical basis of life," (1868), in *Lay sermons, addresses and reviews* (New York, 1883), p. 129.

9. T. H. Huxley, "The Cell Theory," (1853), in *The scientific memoirs of T. H. Huxley*, ed. M. Foster and E. Lankester (London, 1898), vol. 1, p. 277.

10. Ibid., p. 277.

11. T. H. Huxley, "On some organisms living at great depths in the North Atlantic Ocean," *Quart. J. Micro. Sci.* 8 (1868): 203. For a detailed account of the *Bathybius* episode, see P. F. Rehbock, "Huxley, Haeckel and the oceanographers: The case of *Bathybius haeckelii*," *Isis* 66 (1975): 504-33.

12. Huxley, "On some organisms," p. 210.

13. Serialized in *Quart. J. Micro. Sci.* 9 (1869): 27-42, 115-34, 219-32, 327-42.

14. T. H. Huxley, "Biogenesis and abiogenesis," in *Collected essays*, vol. 8, *Discourses biological and geological* (New York, 1894), p. 255.

15. J. Browning, "On the correlation of microscopic physiology and microscopic physics," *Monthly Micro. J.* 2 (1869): 18.

16. E. Haeckel, *Die Radiolarien (Rhizopoda radiaria)* (Berlin, 1862), pp. 231-32n.

17. Quotation from W. Bölsche, *Haeckel, his life and work*, English trans. J. McCabe (London, 1906), p. 134.

18. E. Haeckel, *Generelle Morphologie der Organismen ...* (Berlin, 1886), vol. 2, p. 451.

19. E. Haeckel, *Freie Wissenschaft und Freie Lehre* (1878), English trans. T. H. Huxley, as *Freedom in science and teaching* (New York, 1879), p. 2.

20. Haeckel, *Generelle Morphologie*, vol. 1, p. 164.

21. F. Ballard, *Haeckel's Monism false* (London, 1905), p. 99.

22. Haeckel, *History of creation*, vol. 1, p. 343.

23. E. Haeckel, "Monograph on Monera," *Quart. J. Micro. Sci.* 9 (1869): 30. That part of chapter five dealing with the work of Ernst Haeckel is derived from my article, "The spontaneous generation controversy (1859-1880): British and German reactions to the problem of abiogenesis," *J. Hist. Biol.* 5 (1972): 285-319.

24. Haeckel, *History of creation*, vol. 1, p. 342.

25. H. von Helmholtz, "Über die Erhaltung der Kraft" (1847), in *Ostwald's Klassiker der Exakten Wissenschaften*, No. 1. (Leipzig, 1889), p. 4.

26. T. Schwann, *Microscopical researches into the accordance in the structure and growth of animals and plants,* trans. H. Smith (London, 1847), p. 187.

27. H. von Helmholtz, "Aim and progress of physical science" (1869), in *Popular lectures on scientific subjects,* trans. E. Atkinson (New York, 1873), p. 385.

28. E. du Bois-Reymond, "Darwin versus Galiani" (1876), in *Reden von du Bois-Reymond* (Leipzig, 1886), vol. 1, p. 216.

29. H. Richter, "Zur Darwin'schen Lehre," *Jahrb. inausland ges. Med.* 126 (1865): 243–49.

30. F. A. Lange, *The history of materialism and criticism of its present importance* (1865), English trans. E. C. Thomas, 3rd ed. (New York: Harcourt, Brace, 1925), sect. 2, bk. 2, chap. 3, p. 23.

31. Haeckel, *History of creation,* vol. 1, p. 97.

32. E. du Bois-Reymond, "Über die Grenzen des Naturkennens" (1872), in *Reden von du Bois-Reymond* vol. 1, pp. 105–40.

33. M. J. S. Rudwick, "Uniformity and progression: Reflections on the structure of geological theory in the age of Lyell," in *Perspectives in the history of science and technology,* ed. D. Roller (Norman: University of Oklahoma Press, 1971), p. 220.

34. E. de Beaumont, "Recherches sur quelque-unes des révolutions de la surface du globe," *Annales des sci. naturelles* 18 (1829): 5–25, 284–416; 19 (1830): 5–99, 177–240. See Rudwick, "Uniformity and progression," p. 219.

35. The classical argument for the strong theological element in British geology is given in C. Gillispie, *Genesis and geology* (Cambridge, Mass.: Harvard University Press, 1951). Recent scholarship now suggests this element has been exaggerated. See M. J. S. Rudwick *The meaning of fossils* (London: MacDonald and Co., 1972), and "Uniformity and progression."

36. G. Cuvier, *Essay on the theory of the earth,* English trans. R. Kerr (Edinburgh, 1813), p. 125.

37. P. Flourens, "Éloge historique de Georges Cuvier," in G. Cuvier, *Discours sur les révolutions du globe* (Paris, 1850), p. xxv.

38. W. Buckland, *Geology and mineralogy considered with reference to natural theology, Bridgewater Treatises,* vol. 6 (Philadelphia, 1837).

39. Charles Lyell, *Principles of Geology,* 1st ed. 1830 (London: Johnson Reprint, 1969), vol. 1, chap. 1. Also M. J. S. Rudwick, "The strategy of Lyell's *Principles of Geology,*" *Isis* 61 (1970): 5–33.

40. T. H. Huxley, "The Origin of Species" (1860), in *Collected essays,* vol. 2: *Darwiniana* (New York, 1896), p. 54.

41. C. Darwin, *The origin of species by means of natural selection, or the preservation of favoured races in the struggle for life* (London, 1859); facs. of 1st ed. (Cambridge, Mass.: Harvard University Press, 1966), p. 490.

42. C. Darwin, letter to J. D. Hooker, 29 March 1863, in F. Darwin, ed., *The life and letters of Charles Darwin* (London: Johnson Reprint, 1969), vol. 3, p. 17.

43. *The Times* (London), 19 September 1870. This background to nineteenth-century British science has been discussed by A. Ellegård, "The Darwinian theory and nineteenth century philosophies of science," *J. Hist. Ideas* 18 (1957): 362–99.

44. J. S. Mill, *A system of logic,* in E. Nagel, ed., *John Stuart Mill's philosophy of scientific method* (New York: Hafner, 1950), p. 194.

45. J. S. Mill, *Positive philosophy of Auguste Comte* (New York, 1873), p. 16.

46. C. Darwin, letter to V. Carus, 21 November 1866, in F. Darwin, ed., *More letters of Charles Darwin* (London, 1903), vol. 1, p. 273.

47. Darwin, letter to J. D. Hooker, in F. Darwin, *Life and letters,* vol. 3, p. 17.

48. Darwin, letter to C. Lyell, 18 February 1860, in F. Darwin, *More letters,* vol. 1, p. 140.

49. W. M. Frazer, *A history of English public health, 1834–1939* (London: Bailliere, Tindall and Cox, 1950); N. Longmate, *King Cholera; The biography of a disease* (London: Hamish Hamilton, 1966).

50. For an interesting discussion of quarantine laws, see E. Ackerkneckt, "Anticontagionism between 1821 and 1867," *Bull. Hist. Med.* 22 (1948): 562–93. Although sup-

port for miasmatic concepts continued, the classic work of J. Snow in 1849 and 1855 led the British to associate cholera with water-borne contagia, particularly after the 1865 epidemic.

51. J. Lister, "Address to British Medical Association," *Brit. Med. J.*, 26 August 1871, p. 227.

52. L. S. Beale, *Disease germs, their nature and origin* (London, 1872), p. 9.

53. C. Darwin, *The variation of animals and plants under domestication*, 2 vols. (New York, 1868), p. 448.

54. Ibid., p. 452n.

55. Ibid., p. 454.

56. J. Ross, "The graft theory of disease," *Brit. Med. J.*, 12 October 1872, p. 409.

57. Ibid., p. 410.

58. J. Ross, *Germinal matter and the contact theory* (London, 1867), p. 24.

59. B. W. Richardson, "The germ theory of disease," *Brit. Med. J.* 2 (1870): 467.

60. B. W. Richardson, "A theory as to the natural or glandular origin of contagious diseases," *Nature* 16 (1877): 481.

61. B. W. Richardson, "On the theory and mode of propagation of cholera," *Trans. Epid. Soc.* 2 (1866): 431.

62. J. B. Sanderson, "Lectures on the occurrence of organic forms in connection with contagious and infective diseases," *Brit. Med. J.*, 13 February 1875, p. 200.

63. J. Tyndall, "Spontaneous generation" (1878), in *Essays on floating—matter of the air, in relation to putrefaction and infection* (New York, 1882), p. vii.

64. J. Tyndall, "Dust and Disease," lecture to Royal Institute, *Brit. Med. J.*, 24 June 1871, p. 661.

65. H. C. Bastian, "Remarks on heterogenesis in its relation to certain parasitic diseases," *Brit. Med. J.* (1872): 417.

66. H. C. Bastian, "Discussions on the germ theory of disease," *Trans. Path. Soc. London* 26 (1875): 283.

67. W. B. Carpenter, "On the mutual relations of the vital and physical forces," *Phil. Trans. Roy. Soc.* (1850): 727.

68. Ibid., p. 752.

69. G. L. Geison, "The protoplasmic theory of life and the vitalist-mechanist debate," *Isis* 60 (1969): 285.

70. L. S. Beale, *The mystery of life: An essay in reply to Dr. Gull's attack on the theory of vitality* (London, 1871), p. 10.

71. Ibid., p. 17.

72. Ibid., p. 2.

73. J. H. Bennett, "On the molecular theory of organization," *Proc. Roy. Soc. Edinburgh* 4 (1862): 436–46. In this paper he wrote, "neither does it give any countenance to the doctrines of equivocal or spontaneous generation. It is not a fortuitous concourse of molecules that can give rise to a plant or animal, but only such a molecular mass as descends from parents and receives the appropriate stimulus to act in certain directions."

74. J. H. Bennett, "On the molecular origin of infusoria," *Pop. Sci. Review* 8 (1869): 52.

75. Ibid., p. 65.

76. E. Parfitt, "Experiments on spontaneous generation," *Monthly Micro. J.* 2 (1869): 253–68.

SIX. THE FRENCH *COUP DE GRÂCE*

1. F. Pouchet, "Note sur des proto-organismes végétaux et animaux, nés spontanément dans l'air artificiel et dans le gaz oxygène," *Comptes-rendus Acad. Sci.* 47 (1858): 979–84.

2. H. Milne-Edwards, "Remarques sur la valeur des faits qui sont considérées par quelques naturalistes comme étant propres à prouver l'existence de la génération spontanée des animaux," *Comptes-rendus Acad. Sci.* 48 (1859): 29.

3. Ibid., p. 24.

4. The series of articles by P.-J. van Beneden, C. Bernard, J.-B. Dumas, H. F. Gaultier de Claubry, E. Jobard, F.-J. Lacaze-Duthiers, A. Payen, and J. L. Quatrefages appeared in *Comptes-rendus Acad. Sci.* 48 (1859).

5. F. Pouchet, "Corps organisés recueillis dans l'air par les flocons de la neige," *Comptes-rendus Acad. Sci.* 50 (1860): 534.

6. F. Pouchet, "Recherches sur les corps introduits par l'air dans les organes respiratoires des animaux," ibid., p. 1121.

7. F. Pouchet, "Remarques sur les objections relatives aux protoorganismes rencontrés dans l'oxygène et l'air artificiel," *Comptes-rendus Acad. Sci.* 48 (1859): 149.

8. F.-R. Châteaubriand, *Génie du christianisme* (Paris, 1802).

9. The summary of the political and religious controversies of the Second Empire is based on the following texts: D. G. Charlton, *Secular religions in France, 1815–1870* (London: Oxford University Press, 1963); A. Dansette, *Religious history of modern France,* English trans. J. Dingle (New York: Herder and Herder, 1961); A. L. Guerard, *French prophets of yesterday: A study of religious thought under the Second Empire* (New York, 1920); W. M. Simon, *European positivism in the 19th century* (Ithaca, N.Y.: Cornell University Press, 1963); J. M. Thompson, *Louis Napoleon and the Second Empire* (London: Blackwell, 1954); Gordon Wright, *France in Modern Times* (Chicago: Rand McNally, 1966).

10. Quotation from Dansette, *Religious history,* p. 275.

11. Ibid., p. 311.

12. C. Royer, *De l'origine des espèces par seléction naturelle,* 2nd ed. (Paris, 1866), p. xxvii. The same preface occurs in the first edition of 1862. Both E. Jourdry, *La Philosophie positive* (Paris, 1872), p. 59, and H. de Valroger, *La genèse des espèces. Études philosophiques et religieuses sur l'histoire naturelle et les naturalistes contemporains* (Paris, 1873), maintained that much opposition to Darwin was generated by Royer's preface.

13. For a discussion of early reaction to Darwin, see, J. Farley, "The initial reactions of French biologists to Darwin's *Origin of Species," J. Hist. Biol.* 7 (1974): 275–300.

14. E. Renan, *The life of Jesus,* English trans. (London, 1863), p. 296.

15. F. Pouchet, *Théorie positive de l'ovulation spontanée et de la fécondation des mammifères et de l'espèce humaine* (Paris, 1847), p. 35.

16. Ibid., p. 33.

17. F. Pouchet, "Note sur les organes digestifs et circulatoires des animaux infusoires," *Comptes-rendus Acad. Sci* 28 (1848): 516–18.

18. F. Pouchet, *Hétérogenie ou traité de la génération spontanée* (Paris, 1859), p. 9.

19. Ibid., p. 7.

20. Ibid., p. 9.

21. Ibid., p. 95.

22. Ibid., p. 97.

23. Ibid., p. 98.

24. Ibid., p. 127.

25. F. J. Pictet, *Traité de paléontologie ou histoire naturelle des animaux fossiles,* 2nd ed. (Paris, 1853), p. 93.

26. Pouchet, *Hétérogenie,* p. 462. For details of the geological debate, see chapter five.

27. Ibid., p. 540.

28. Ibid., p. 588.

29. Ibid., p. vii.

30. J. Tyndall. "Spontaneous generation," *The Nineteenth Century* 3 (1878): 26.

31. W. Bulloch, *The history of bacteriology* (London: Oxford University Press, 1938), p. 92.

32. J. Farley and G. Geison, "Science, politics and spontaneous generation in 19th century France: The Pasteur-Pouchet debate," *Bull. Hist. Med.* 48 (1974): 161–98.

33. Tyndall, "Spontaneous Generation."

34. T. S. Hall. *Ideas of life and matter* (Chicago: University of Chicago Press, 1969), vol. 2, p. 294.

35. A delightful account of Pasteur's crystallography studies leading to his work on fermentation is given in D. Huber, "Louis Pasteur and molecular dissymmetry, 1844–1857" (Master's thesis, Johns Hopkins University, 1969). I am greatly obliged to Ms. Kottler (née Huber) for allowing me to read her thesis.

36. L. Pasteur, "Recherches sur la dissymétrie moléculaire des produits organiques naturels," *Oeuvres de Pasteur*, ed. Pasteur Vallery-Radot (Paris: Masson et Cie, 1922), vol. 1, p. 333. Hereafter this work will be referred to as *O.P.*

37. L. Pasteur, "Mémoire sur la fermentation appelée lactique," *O.P.*, vol. 2. p. 3.

38. Ibid., p. 9.

39. L. Pasteur, "Mémoire sur les corpuscules organisés qui existent dans l'atmosphère: Examen de la doctrine des générations spontanées," *O.P.*, vol. 2, p. 224.

40. L. Pasteur, "Nouveaux faits pour servir à l'histoire de la levure lactique," *O.P.*, vol. 2, p. 36.

41. Lettre manuscrite de Pasteur à Pouchet, 28 February 1859, *O.P.*, vol. 2, p. 628.

42. Pasteur, *O.P.*, vol. 2, p. 246.

43. Ibid., p. 235.

44. Ibid., p. 246.

45. N. Joly and Ch. Musset, "Étude microscopique de l'air," *Comptes-rendus Acad. Sci.* 50 (1860): 748.

46. Pasteur, *O.P.*, vol. 2, p. 274.

47. Ibid., p. 277.

48. G. Pennetier, *Un débat scientifique: Pouchet et Pasteur* (Paris, 1907).

49. F. Pouchet, N. Joly, et Ch. Musset, "Expériences sur l'hétérogenie exécutées dans l'intérieur des glaciers de la Maladetta (Pyrénées)," *Comptes-rendus Acad. Sci.* 57 (1863): 558.

50. L. Pasteur, "Examen du rôle attribué au gaz oxygène atmosphérique dans la destruction des matières animales et végétales après la mort," *O.P.*, vol. 2, p. 170.

51. Pasteur, "Des générations spontanées," *O.P.*, vol. 2, p. 346.

52. F. Guizot, *L'église et la société chrétiennes en 1861* (Paris, 1862), p. 18.

53. E. Faivre, "La question des générations spontanées," *Mémoires L'Acad. Sci., Belles-Lettres et Arts, Lyon* 10 (1860): 170–71.

54. Pasteur, *O.P.*, vol. 2, p. 328.

55. Ibid., p. 328.

56. Ibid., pp. 332–33.

57. These statements were made on at least two separate occasions, in *O.P.*, vol. 2., pp. 295, 334.

58. Quoted in R. Vallery-Radot, *The life of Pasteur* (New York: Dover, 1960), p. 106.

59. P. Flourens, *Examen du livre de M. Darwin sur l'origine des espèces* (Paris, 1864), p. 170.

60. H. Milne-Edwards, *Leçons sur la physiologie et l'anatomie comparée de l'homme et des animaux* (Paris, 1863), vol. 8, leçon 71, p. 270.

61. N. Joly, "Sur l'hétérogenie," *Revue des Corps Scientifiques* 2 (1865): 226.

62. E. Duclaux, *Pasteur: The history of a mind*. English trans. E. Smith and F. Hedges (Philadelphia: W. B. Saunders, 1920), p. 107.

63. G. Pennetier, "Les générations spontanées," *Revue des Corps Scientifiques* 1 (1864): 265.

64. F. Pouchet, N. Joly, and Ch. Musset, "Note adressée par MM. Pouchet. Joly et Musset à la commission des expériences relatives à la génération spontanée," *Revue des Corps Scientifiques* 1 (1864): 431.

65. Ibid., p. 431.

66. É. Alglave, "La question des générations spontanées devant l'Académie des Sciences," *Revue des Corps Scientifiques* 2 (1865): 206.

67. L. Pasteur, "Discussion sur la fermentation" (1875), *O.P.*, vol. 6, p. 54.

68. L. Pasteur, "Note en réponse à des observations critiques présentées a l'Académie par M. M. Pouchet, Joly et Musset," *O.P.,* vol. 2., p. 322.

69. E. Fremy, "Sur les corps hémiorganisés," *Comptes-rendus Acad. Sci.* 58 (1864): 1165.

70. Ibid., p. 1166.

71. E. Fremy, *Comptes-rendus Acad. Sci.* 73 (1871): 1425.

72. L. Pasteur, "Discussion avec MM. Fremy et Trécul sur l'origine et la nature des ferments," *O.P.,* vol. 2, pp. 367–68.

73. A. Trécul, "Recherches sur l'origine des levures lactiques et alcoolique," *Comptes-rendus Acad. Sci.* 73 (1871): 1454.

74. Ibid., p. 1457.

75. L. Pasteur, "Observations à propos d'une note de M. Trécul," *O.P.,* vol. 2, p. 369.

76. L. Pasteur, "Nouvelles expériences pour démontrer que le germe de la levure qui fait le vin provient de l'extérieur des grains de raisin," *O.P.,* vol. 2, p. 386.

77. L. Pasteur, *Études sur la bière,* English trans. F. Faulkner and D. Robb as *Studies on fermentation: The diseases of beer, their causes, and the means of preventing them* (London, 1879), p. 152.

78. C. Bernard, "La fermentation alcoolique," *Revue des cours scientifiques,* 20 July 1878, p. 56. An account of this episode is given by J. M. D. Olmsted, "Claude Bernard's posthumously published attack on Pasteur and Pasteur's defense," *Ann. Med. Hist.* 9 (1937): 114–24.

79. L. Pasteur, "Examen critique d'un écrit posthume de Claude Bernard sur la fermentation alcoolique," *Comptes-rendus Acad. Sci.* 87 (1878): 815.

80. Ibid., p. 819.

81. A Trécul, "Remarques sur l'origine des levures, lactiques et alcoolique," *Comptes-rendus Acad. Sci.* 75 (1872): 1163.

82. Farley and Geison, "Science, politics and spontaneous generation."

83. L. Pasteur, "La dissymétrie moléculaire" (1883), *O.P.,* vol. 1, p. 375.

84. Ibid., p. 377.

85. F. Pouchet, *Les Créations successives et les soulevèments du globe: Lettres à M. Jules Desnoyers* (Paris, 1862), p. 10.

86. F. Pouchet, *Nouvelles expériences sur la générations spontanée et la resistance vitale* (Paris, 1864), chap. 5, p. 199.

87. G. Pennetier, *L'Origine de la vie* (Paris, 1868), p. vii.

88. Pasteur, *Études sur la bière, O.P.,* vol. 5, p. 79.

89. Pasteur, *O.P.,* vol. 2, p. 295.

90. Pasteur, "Discussion sur la fermentation," *O.P.,* vol. 6, p. 57.

91. Pasteur, *O.P,* vol. 1, p. 375.

92. Pasteur, "Observations sur les forces dissymétriques," *O.P,* vol. 1, p. 362.

93. Pasteur, "Sur l'origine de la vie," *O.P,* vol. 7, p. 30.

94. "La dissymétrie moléculaire," *Revues Scientifiques* (1885): 65.

95. F. Isnard, *Spiritualisme et matérialisme* (Paris, 1879), p. viii.

96. Ibid., p. 85.

97. Ibid., p. 86.

98. E. Perrier, "Le transformisme et les sciences physiques," *Revues Scientifiques* (1879): 890–95.

99. "Hypothèses récentes sur l'origine cosmique de la vie," *Revues Scientifiques* (1882): 352.

100. E. Littre, "L'hypothèse de la génération spontanée et celle du transformisme," *La Phil. Positive* 22 (1879): 172.

101. Ibid., p. 166.

102. A Gaudry, *Les Ancêtres de nos animaux dans le temps géologiques* (Paris, 1888), p. 32.

103. De Valroger, *La genèse des espèces,* p. 57.

104. P. E. Chauffard, "La Science et l'ordre social," in *La Vie, études et problèmes de biologie générale* (Paris, 1878), p. 461.

105. Ibid., p. 486.
106. Quoted by N. Joly, "Sur l'hétérogenie," *Revue des Corps Scientifiques* 2 (1865): 226.

SEVEN. THE LAW OF GENETIC CONTINUITY

1. H.C. Bastian, "Facts and reasonings concerning the heterogenous evolution of living things," *Nature* 2 (1870): 171.
2. Ibid., p. 172.
3. Ibid., p. 228.
4. H. N. Moseley, Review of Bastian's *The beginnings of life,* in *The Academy* 3 (1872): 411. This point is stressed in G. Vandervliet, *Microbiology and the spontaneous generation debate during the 1870s* (Lawrence, Kansas: Coronado Press, 1971).
5. H. C. Bastian, "Remarks on heterogenesis in its relation to certain parasitic diseases," *Brit. Med. Journal* 1 (1872): 418.
6. F. E. Abbot, "A review of Spencer's *The principles of biology,*" in *North American Review* 221 (1868): 386.
7. Ibid., p. 393.
8. G. W. Child, "Recent researches on the production of the lowest forms of animal and vegetable life" (1864), in *Essays on physiological subjects* 2nd ed. (London, 1869), p. 67.
9. Child, "Some aspects of the theory of evolution," in *Essays,* p. 140.
10. Ibid., p. 144.
11. H. C. Bastian, "A review of J. H. Stirling's, As regards protoplasm, in relation to Prof. Huxley's essay on the physical basis of life," *Nature* 1 (24 February 1870): 424.
12. H. C. Bastian, *The beginnings of life, being some account of the nature, modes of origin and transformations of lower organisms,* 2 vols. (London, 1872), p. 79.
13. Ibid., chap. 5, "Origin of reproductive cells."
14. Bastian, "Remarks on heterogenesis," p. 201.
15. H. C. Bastian, *Evolution and the origin of life* (London, 1874), p. 39.
16. Ibid., p. 184.
17. Ibid., pp. 4 and 15.
18. A. Wallace, "*The beginnings of life,* a review of Bastian's book," *Nature* 6 (1872): 303. Wallace even remarked that Bastian's book would have ranked with the *Origin of species* had it been written in a more lucid style.
19. T. R. R. Stebbing, "Note on the hypothesis of spontaneous generation," in *Essays on Darwinism* (London, 1871), p. 127. For details on the life and work of Stebbing, see E. L. Mills, "Amphipods and equipoise: A study of T. R. R. Stebbing," in *Growth by intussusception: Ecological essays in honour of G. Evelyn Hutchinson,* ed. E. Deevey, in *Trans. Connecticut Acad. Arts and Science* 44 (1972): 237–56.
20. Wallace, "Review of Bastian's *Beginnings of life,*" p. 285.
21. T. H. Huxley, "Biogenesis and abiogenesis" (1870), in *Discourses biological and geological* (New York, 1894), p. 255.
22. Ibid., p. 259.
23. W. B. Carpenter, *A Manual of physiology including physiological anatomy,* 4th ed. (London, 1865), p. 44.
24. J. H. Stirling, *As regards protoplasm, in relation to Prof. Huxley's essay on the physical basis of life* (New Haven, 1870), p. 64.
25. L. S. Beale, *The mystery of life: An essay in reply to Dr. Gull's attack on the theory of vitality* (London, 1871), p. 42.
26. Ibid., p. 32.
27. J. B. Pettigrew, "A lecture on the relations of plants and animals to inorganic matter and of the interaction of the vital physical forces," *Lancet* 2 (1873): 691–96.
28. W. Thiselton-Dyer, "On spontaneous generation and evolution," *Quart. J. Micro. Sci* 10 (1870): 335.

209

29. Ibid., 354.

30. H. Spencer, *Spontaneous generation and the hypothesis of physiological units: A reply to the North American Review* (New York, 1870), p. 4; reprinted as appendix to vol. 1 of his later editions of *Principles of biology,* including a New York edition of 1898.

31. J. B. Sanderson, "Dr. Bastian's experiments on the beginning of life," *Nature* 7 (1873): 180.

32. G. Bentham, "Anniversary address to the Linnean Society," *Nature* 6 (1872): 132.

33. W. Roberts, "Dr. Bastian's experiments on the beginning of life," *Nature* 7 (1873): 302.

34. J. B. Sanderson, "Dr. Bastian's turnip-cheese experiments," *Nature* 8 (1873): 141.

35. J. B. Sanderson, "Dr. Bastian's turnip-cheese experiments," *Nature* 8 (1873): 181.

36. H. C. Bastian, "On the temperature at which bacteria, vibriones, and their supposed germs are killed when immersed in fluids or exposed to heat in a moist state," *Nature* 7 (1873): 413.

37. E. R. Lankester, "Dr. Sanderson's experiments," *Nature* 7 (1873): 242.

38. J. Lister, "On the germ theory of putrefaction and other fermentative changes," *Nature* 8 (1873): 212, 232.

39. F. Cohn, "Untersuchungen über Bacterien," in *Beiträge zur Biologie der Pflanzen* 1 (1872): 127–224; partial and incomplete review in *Nature* 7 (1873): 300.

40. J. Tyndall, "On the optical deportment of the atmosphere in reference to the phenomena of putrefaction and infection," *Proc. Roy. Soc.* 24 (1876): 171–83; Abstract in *Nature* 13 (1876): 252–54, 268–70.

A parallel series of experiments took place in Germany at this time, in which the same basic questions were being posed, particularly whether bacterial germs (*Keime*) were more heat resistant than the bacteria. As in Britain, the experiments were initiated by Sanderson's support of Bastian and also by the claims of D. Huizinga, professor of physiology at Groningen in Holland, who discussed the issue between 1873 and 1875. He had generated bacteria from a mixture of salt, ammonium tartrate, grape sugar, and peptone and claimed that heating at 100°C for 5 to 10 minutes was sufficient not only to kill the bacteria, but also to "destroy the germinating capability of their hypothetical germs." "Zur Abiogenesis-Frage," *Archiv. f. gesammte Physiol. d. Menschen u. Thiere* 7 (1873): 556. He was opposed by Paul Samuelson of Königsberg, Richard Gscheidlen of Breslau, and Felix Putzeys of Lüttich. Samuelson seemed to have been more aware of the problems in these experiments than anyone else. He wrote of his own experiments, for example: "These experiments are not and cannot be a disproof of spontaneous generation in general, but only a disproof of the possibility of abiogenesis in the experimental solution under consideration. We can only show that the origin of organisms without preexisting germs is not proven in this case." "Über Abiogenesis," Ibid., 8 (1873): 288. For details of the German debate, see Ibid., 8 (1873): 180, 277, 551; 9 (1874): 163, 391; 10 (1875): 62; 11 (1875): 387.

41. J. Tyndall, "On floating matter and beams of light," *Nature* 1 (1870): 501.

42. Tyndall, "Optical deportment of the atmosphere," p. 173.

43. J. Tyndall, "Spontaneous generation: A last word," *The Nineteenth Century* 3 (1878): 501, 507.

44. J. Tyndall, "The Belfast Address," in *Fragments of science, A series of detached essays, addresses, and reviews* (New York, 1892), p. 191.

45. Ibid., p. 192.

46. J. Tyndall, "Introductory note," (1881), *Essays on the floating-matter in the air, in relation to putrefaction and infection* (New York, 1882), p. vii.

47. "Inquirer," "Prof. Tyndall on germs," *Nature* 13 (1876): 347.

48. F. Cohn, "Untersuchungen über Bacterien II," in *Beiträge zur Biologie der Pflanzen* I, no. 3 (1875): 141–207; reviewed in *Nature* 14 (1876): 202.

49. F. Cohn, "Untersuchungen über Bacterien IV," in *Beiträge zur Biologie der Pflanzen* II, no. 2 (1876): 249–76.

50. J. Tyndall, "Fermentation and its bearing on surgery and medicine" (1876), in *Fragments of science*, vol. 2, pp. 251-89.

51. J. Tyndall, "Preliminary note on the development of organisms in organic infusions," *Proc. Roy. Soc. London* 25 (1877): 504.

52. H. C. Bastian, "On the conditions favouring fermentation and the appearance of Bacilli, Micrococci and Torulae in previously boiled liquids," *J. Linn. Soc (Zool.) London* 14 (1877): 71.

53. Ibid., p. 80.

54. E. R. Lankester, "Ernst Haeckel on the mechanical theory of life and on spontaneous generation," *Nature* 3 (1871): 354-56.

55. J. Y. Buchanan, "Preliminary report to Prof. Wyville Thomson on work done on board H.M.S. *Challenger*," *Proc. Roy. Soc. London* 24 (1876): 605.

H.M.S. *Challenger*, a 2306-ton corvette, sailed from England in December 1872, returning three and one-half years later. The expeditions basic tasks were to determine the physical characteristics of the deep sea and the chemical composition of sea water, to examine marine sediments, and to study the distribution of marine organisms. To that end the *Challenger* circumnavigated the globe, making over 300 stations. In the words of Eric Mills, the voyage represented "the first great quantum jump in understanding the oceans." E. Mills, "The *Challenger* expedition: How it started and what it did," in *One hundred years of oceanography* (Halifax: Dalhousie University Press, 1975).

56. T. H. Huxley, "Notes from the *Challenger*," *Nature* 12 (1875): 315.

57. H. C. Bastian, "Researches illustrative of the physicochemical theory of fermentation and of the conditions favouring archebiosis in previously boiled fluids," *Nature* 14 (1876): 220-22.

58. H. C. Bastian, "The fermentation of urine and the germ theory," *Nature* 14 (1876): 309-11.

59. Ibid., p. 309.

60. W. Roberts, "The spontaneous generation question, Note on the influence of liquor potassae and an elevated temperature on the origin and growth of microphytes," and J. Tyndall, "Note on the deportment of alkalized urine," *Nature* 15 (1877): 302.

61. L. Pasteur and J. F. Joubert, "Note sur l'altération de l'urine," in *Nature* 15 (1877): 314.

62. Reply by H. C. Bastian in *Nature* 15 (1877): 314.

63. Reply by L. Pasteur in *Nature* 15 (1877): 380.

64. A detailed account of this episode was presented by H. C. Bastian in *Nature* 16 (1877): 276-79, "The Commission of the French Academy and the Pasteur-Bastian Experiments." My account is taken directly from this report. Note that the commission's members were those who had previously behaved so atrociousely towards Pouchet, Joly, and Musset in 1864.

65. J. Tyndall, "On heat as a germicide when discontinously applied," *Proc. Roy. Soc.* 25 (1877): 569.

66. J. Tyndall, "Spontaneous generation," *The Nineteenth Century* 3 (1878): 45.

67. Ibid., pp. 36-37.

68. H. C. Bastian, "Spontaneous generation: A reply," *The Nineteenth Century* 3 (1878): 265.

69. Ibid., p. 267.

70. Ibid., p. 274. None other than Lister denied the existence of ultramicroscopic germs and the ability of any organism to resist boiling temperatures; *Brit. Med. J.* (1877): 902.

71. Tyndall, "Spontaneous generation: A last word," p. 507.

72. T. H. Huxley, Prefatory note to his translation of E. Haeckel's *Freedom in science and teaching* (New York, 1879), p. xii.

73. J. Tyndall, "Prof. Virchow and evolution," in *Fragments of science*, vol. 2, pp. 402-4.

74. E. du Bois-Reymond, "Darwin versus Galiani," in *Reden von du Bois-Reymond* (Leipzig, 1886), vol. 1, pp. 211-39.

75. J. Müller, "Upon the development of molluscs in Holothuriae," *Ann. et Mag. Nat. Hist.* 9 (1852): 22-37. Curiously, Müller denied that such events were analogous to

"the theory of equivocal generation of intestinal worms, which has long since taken its place in the category of errors." He claimed instead that it represented another example of "heterogony," or the production of beings dissimilar to the parent. Thus the production of organisms of one species from living material of another, once seen as illustrating the validity of spontaneous generation, could also be seen in terms of Steenstrup's alternation of generations. As Albert von Kölliker remarked in 1864, "if the vermiform trematode 'nurse' can develop within itself the very unlike cercaria, it will not appear impossible that the egg, or ciliated embryo, of a sponge, for once, under special conditions, might become a hydroid polyp, or the embryo of the medusa, an echinoderm." "Über die Darwin'sche Schöpfungstheorie," *Zeit f. wissen. Zool.* 14 (1864): 184. On the other hand, Büchner claimed support for spontaneous generation from Steenstrup's concept.

76. H. von Helmholtz, "The origin of the planetary system" (1871), in *Selected writings of Hermann von Helmholtz*, ed. R. Kahl (Middletown, Conn.: Wesleyan University Press, 1971), p. 294.

77. W. Thomson and P. G. Tait, *Handbuch der theoretischen Physik*, German trans. H. von Helmholtz and G. Wertheim (Brunswick), vol. 2 (1874), p. xi. Also see note 27, in chapter five.

78. A. Weismann, "The duration of life" (1881), in *Essays upon hereditary and kindred biological problems*, English trans. by Poulton et al. (Oxford: Clarendon Press, 1889), p. 33.

79. C. von Nägeli, *Mechanisch-Physiologische Theorie der Abstammungslehre* (Munich, 1884), p. 83.

80. R. Virchow, *"The freedom of science in the modern state"* (London, 1878), p. 64.

81. Ibid., pp. 32, 36.

82. Ibid., p. 39.

83. L. S. Beale, *The new materialism: Dictatorial scientific utterances and the decline of thought* (London, 1883), p. 30.

84. Quoted in R. Vallery-Radot, *The life of Pasteur* (New York: Dover, 1960), p. 256.

85. L. Pasteur, J. F. Joubert, and C. E. Chamberland, "La théorie des germes et ses applications à la médicine et à la chirurgie," *O.P.*, vol. 6, p. 116.

86. R. Koch, "The etiology of tuberculosis" (1882), in *Recent essays on bacteria in relation to disease*, ed. W. Cheyne (London, 1886), p. 67.

87. R. Koch, "The etiology of cholera" (1884), in Cheyne, *Recent Essays*, p. 360.

88. C. von Nägeli, "The lower fungi in relation to infectious disease" (1877), in T. Smith, "Koch's views on the stability of species among bacteria," *Ann. Med. Hist.* 4 (1932): 524–30.

89. R. Koch, "On the investigation of pathogenic organisms" (1881), in Cheyne, *Recent essays*, p. 39.

90. An exhaustive bibliography of these cytological investigations is given in E. L. Mark, "Maturation, fecundation and segmentation of *Limax campestris*, Binney," *Bull. Mus. Comp. Zool. Harvard* 6 (1881): 173–625. The best résumé of these findings is certainly the one in W. Coleman, "Cell, nucleus and inheritance: An historical study," *Proc. Amer. Phil. Soc.* 109 (1965): 124–58. Much of my brief account is taken from Coleman's study.

91. C. Gegenbaur, *Grundriss der vergleichenden Anatomie* (Leipzig, 1874), p. 17. J. Sachs, *Textbook of botany*, English trans. A. W. Bennett and W. T. Thiselton-Dyer (Oxford, 1875), contains a detailed description of free cell-formation in plants on page 11.

92. See A. Hughes, *A history of cytology* (New York: Abelard-Schuman, 1961); and S. Bradbury, *The evolution of the microscope* (Oxford: Pergamon Press, 1968).

93. F. Balfour, *A monograph on the development of elasmobranch fishes* (London, 1878).

94. W. Flemming, "Beiträge zur Kenntnis der Zelle und ihrer Lebenserscheinungen: Theil I," *Arch. Mikro. Anat.* 16 (1878): 302–436.

95. W. Flemming, *Zellsubstanz, Kern und Zelltheilung* (Leipzig, 1882).

96. E. Strasburger, "Die Controversen der indirecten Kerntheilung," *Arch. Mikro. Anat.* 23 (1884).

97. Coleman, "Cell, nucleus and inheritance," p. 136.

98. A more detailed discussion of the "contact" theory is given by Coleman, ibid., pp. 133–36.

99. O. Hertwig, "Beiträge zur Kenntnis der Bildung, Befruchtung und Theilung der thierischen Eies," *Morph. Jahrb* 1 (1876): 347–434.

100. E. van Beneden, "Contributions to the history of the germinal vesicle and of the first embryonic nucleus," *Quart. J. Micro. Sci.* 16 (1876): 154.

101. H. Fol, "Sur le commencement de l'hénogénie chez divers animaux," *Arch. Zool. Expér. et gén.* 6 (1877): 145–69.

102. O. Hertwig, *Die Zelle und die Gewebe* (Jena, 1893), p. 187.

103. Ibid., p. 145.

104. E. B. Wilson, *The cell in development and inheritance*, 1st ed. (New York, 1896), p. 9. In the second edition (1924), this passage was changed slightly and the words, "life under existing conditions never arises *de novo,*" were added on page 11.

105. W. H. Dallinger, "Recent researches into the origin and development of minute and lowly life-forms, with a glance at the bearing of these on the origin of bacteria," *Nature* 16 (1877): 24. The importance of this law of genetic continuity was stressed by J. W. Wilson, "Biology attains maturity in the nineteenth century," in M. Clagett, ed., *Critical problems in the history of science* (Madison: University of Wisconsin Press, 1969), pp. 401–18. In opposition to Wilson, however, I see the establishment of the law to have occurred approximately twenty years after the work of Pasteur, Darwin, and Virchow in the 1850s.

106. H. C. Bastian, *The nature and origin of living matter* (London: Fisher Unwin, 1905), pp. 138, 145.

EIGHT. DEMISE AND RECOVERY

1. E. Ray Lankester, "Presidential address to B.A.A.S.," *Report of the B.A.A.S.* (1906), p. 24.

2. R. Muir and J. Ritchie, *Manual of bacteriology*, 6th ed. (London, 1913).

3. C. O. Whitman, *"Evolution and epigenesis,"* in *Woods Hole Lectures* (Boston, 1895), p. 214.

4. As far as I can ascertain, nineteenth-century neolamarckians paid little or no attention to the spontaneous generation issue. The neolamarckian H. F. Osborn published *The origin and evolution of life* in 1918, but by then "classical" nineteenth-century biology had broken down.

5. H. C. Bastian, *Evolution and the origin of life* (London, 1874), p. 15.

6. W. Preyer, *Die Hypothesen über den Ursprung des Lebens* (Berlin, 1880). Quoted in A. I. Oparin, "The origin of life," appendix to J. D. Bernal, *The origin of life* (London: Weidenfeld and Nicolson, 1967), p. 227.

7. S. Arrhenius, *Lehrbuch der kosmischen Physik* (Leipzig, 1903), and *Worlds in the making: The evolution of the universe* (New York, 1908).

8. F. Lange, *The history of materialism and criticisms of its present importance* (1865), English trans. E. C. Thomas, 3rd ed. (New York: Harcourt, Brace, 1925), sect. 2, bk. 2, chap. 3, p. 19.

9. E. A. Schaefer, "Presidential address to B.A.A.S.," *Report of B.A.A.S.* (1912), p. 13.

10. H. C. Bastian, *The nature and origin of living matter* (London: Fisher Unwin, 1905), p. 145. In these years he published also *Studies in heterogenesis* (1904); *The evolution of life* (1907); and *The origin of life* (1911). There are many indications supporting my contention that these books were virtually ignored. The Harvard libraries house all of Bastian's earlier works, but *none* of the above. Two of them were located in theological colleges: Acadia University in Nova Scotia with strong Baptist affiliations and St. Michael's Roman Catholic College at the University of Toronto. The latter's copy of *The evolution of life* has a small communique by H. Onslow attached, in which it is stated that, "The results obtained indicate that the method employed yields tubes which were absolutely sterile for all periods up to three years." In addition, an old article by

"B.C.A.W." is affixed to the blank leaves of the book, in which he/she not only refers to the book as, "rather ancient history" but goes on to state, "The plain man confronted by such a state of affairs will probably reason that though Athanasius may be right against the world, a good deal of proof will require to be forthcoming before such sweeping assertions as Dr. Bastian makes can be accepted in face of the absolute unanimity of the rest of the scientific world." In the foreword to *The origin of life,* Bastian reported that the book was submitted to the Royal Society in 1910, but was "not considered suitable for acceptance." In the text, on page 13, he remarked, "What is the object of the Society but to Advance Natural knowledge? And how can it expect to do this if it tries to stifle or ignore that which is adverse to generally accepted beliefs?" Bastian was neither the first nor the last to have suffered the inequities of the referee system!

11. Bastian, *Nature and origin of living matter,* pp. 138, 140.

12. Ibid., p. 144.

13. Ibid., p. 147.

14. Ibid., p. 148.

15. Ibid., p. 151.

16. Ibid., p. 315.

17. Ibid., p. 329.

18. Ibid., p. 330.

19. Ibid., p. 329.

20. Ibid., p. 157.

21. M. Planck, *Wissenschaftliche Selbstbiographie* (Leipzig: Johann Barth, 1955), p. 22.

22. B. Moore, *The origin and nature of life* (London: Thornton Butterworth, 1921), p. 189. F. J. Allen spoke earlier of a nitrogen-based "living molecule." *Report of B.A.A.S.* (1896), p. 983, and *Proc. Birmingham Nat. Hist. and Phil. Soc.* 11 (1899): 44–67.

23. Most authors claim that this conclusion was not reached until much later in the twentieth century. However, by the early part of the century, work on stellar evolution and radioactivity suggested that all elements and all compounds were fundamentally aggregates of hydrogen, which slowly built up more and more complex forms as the temperature of the earth decreased. As early as 1889, J. R. Rydberg claimed that proteins eventually were produced in this manner.

24. F. R. Japp, "Stereochemistry and vitalism," *Nature* (September 1898): 459.

25. See *Nature* for September 22, September 29, October 6, November 10, and December 1, 1898.

26. G. B. Kauffman, "The discovery of optical activity co-ordination compounds: A milestone in stereochemistry," *Isis* 66 (1975): 38–62.

27. H. C. Bastian, *The evolution of life* (London, 1907), p. 7.

28. W. G. H. Brehmer, "The origin of the living organism in the light of the new physics," *Medical Times* 55, No. 11 (1927): 247–50, 258.

29. Schaefer, "Presidential address," p. 10.

30. For details of this early work on viruses, see C. E. Simon, "The filterable viruses," *Physiol. Reviews* 3 (1923): 483–508.

31. T. Graham, "Liquid diffusion applied to analysis," *Phil. Trans. Roy. Soc.* 151 (1861): 183–224.

32. M. Florkin, *A history of biochemistry,* in M. Florkin and E. Stoltz, *Comprehensive biochemistry,* vol. 30 (Amsterdam: Elsevier Co., 1972), p. 298.

33. E. Fisher (1907), quoted in J. S. Fruton, *Molecules and life* (New York: Wiley, 1972), p. 135.

34. Ibid., p. 140.

35. This change in approach to proteins is discussed by J. T. Edsall, "Proteins as macromolecules: An essay on the development of the macromolecule concept and some of its vicissitudes," *Archiv. of Biochem. Biophysics,* (Supplement 1) (1962): 12–20.

36. R. E. Kohler, "The enzyme theory and the origin of biochemistry," *Isis* 64 (1973): 181–96.

37. Ibid., p. 185.

38. Ibid., pp. 185–86.

39. V. L. Kellogg, *Darwinism today* (New York, 1907). Hostility to spontaneous generation is very evident: "If such a summary disposal of the theories of spontaneous generation and divine creation is too repugnant to my readers to meet with their toleration, then, . . . my book and such readers had better immediately part company; we do not speak the same language."

40. A. B. Macallum, "On the origin of life on the globe," *Trans. Can. Inst.* 8 (1908): 435–36.

41. Ibid., p. 437. He denied strongly that such events could occur today, claiming an "inconceivably long period of time" was required.

42. L. T. Troland, "The chemical origin and regulation of life," *The Monist* 24 (1914): 103.

43. L. T. Troland, "Biological enigmas and the theory of enzyme action," *Amer. Nat.* 51 (1917): 321–50.

44. E. A. Minchin, "The evolution of the cell," *Amer. Nat.* 50 (1916): 5–39, 106–19.

45. F. D'Hérelle, *The bacteriophage and its behavior,* English trans. G. Smith (Baltimore: Williams and Wilkins, 1926) p. 309.

46. Ibid., p. 316*n.*

47. Ibid., p. 359.

48. Ibid., p. 358.

49. Ibid., p. 377.

50. F. D'Hérelle, *Immunity in natural infectious disease,* trans. G. Smith (Baltimore: Williams and Wilkins, 1924), p. 348.

51. See chapter 14, "The dark age of biocolloidology," in Florkin's *History of biochemistry.*

52. A. I. Kendall, "Bacteria as colloids," in H. N. Holmes, ed., *Colloid symposium monograph* (New York: The Chemical Catalogue Co., 1925), vol. 2, p. 195.

53. V. C. Vaughan, "A chemical concept of the origin and development of life," *Chem. Reviews* 4 (1927): 172.

54. "Life and death," *Nature* 122 (1928): 503.

55. E. C. Baly et al., "Photosynthesis of naturally occurring compounds: I. Action of ultra-violet light on carbonic acid," *Proc. Roy. Soc.* 116 (1927): 197–211.

56. E. Nordenskiold, *The history of biology* (New York: Tudor, 1928), p. 595. After many remarks of this vintage one appreciates W. B. Hardy's of the same year the more: "At present the colloidal kingdom seems to be an Alsatia wherein difficult states of matter find refuge from a too-exacting enquiry," in Holmes, ed., *Colloid symposium monograph,* vol. 6, p. 8.

57. J. Alexander, ed., *Colloid chemistry,* vol. 2, *Biology and medicine* (New York: The Chemical Catalogue Co., 1928).

58. C. E. Simon, "The filterable viruses," in ibid., chap. 27, p. 527.

59. J. Alexander and C. B. Bridges, "Some physico-chemical aspects of life, mutation and evolution," ibid., chap. 1, p. 17.

60. H. J. Muller, "Variation due to change in the individual gene" (1921), and "The gene as the basis of life" (1926) in *Studies in Genetics: The selected papers of H. J. Muller* (Bloomington: Indiana University Press, 1962).

61. S. Leduc, "Solutions and life," and A. L. Herrera, "Plasmogeny," in Alexander, ed., *Colloid chemistry,* vol. 2, chaps. 2 and 3. These strange experiments are discussed at length in A. I. Oparin, *The evolution of life on the earth,* English trans. A. Synge (Edinburgh: Oliver and Boyd, 1957), pp. 86–92, and by R. Beutner, *Life's beginning on the earth* (Baltimore: Williams and Wilkins, 1938).

62. Oparin, "Origin of life," in Bernal, *Origin of life,* p. 211.

63. Ibid., p. 229.

64. Ibid., p. 230–31. Although this work is of significance to questions about the origin of life, as far as the more limited topic of spontaneous generation is concerned, it represents merely a more detailed account of what was then widely believed, namely, that in the past a colloidal lifelike entity arose suddenly by spontaneous generation.

65. J. B. S. Haldane, "The Origin of Life," in Bernal, *Origin of life,* p. 243.

66. Ibid., p. 246.

67. Ibid., p. 246–47.

68. Ibid., p. 248.

69. J. S. Haldane, *The philosophy of a biologist* (London: Oxford University Press, 1935), p. 70.

70. J. S. Haldane, *Mechanism, life and personality* (London, 1914), p. 64.

71. Ibid., p. 100.

72. As will be discussed in chapter nine, the Marxist approach did not become significant until Oparin's work of 1936.

73. J. B. S. Haldane, "Origin of life," in Bernal, *Origin of life*, p. 248.

74. A. Gardner, *Microbes and ultramicrobes* (1935), quoted in A. I. Oparin, *The Origin of life* (1936), trans. S. Margulis (New York: Macmillan, 1938; Dover reprint edition, 1953).

75. Moore, *Origin and nature of life*, p. 190.

76. D. H. Duckworth, "Who discovered bacteriophage?" *Bact. Reviews* 40 (1976): 793–802.

77. A. J. T. Janse, "Continuous archigony and its bearing on evolution and classification," *South African J. Nat. Hist.* 6 (1928): 208.

78. Troland, "Chemical origin and regulation of life," pp. 102, 105, 109.

79. Oparin, "Origin of life," in Bernal, *Origin of life*, p. 228.

80. E. A. Schaefer, in 1912, wrote:

Nor should we expect the spontaneous generation of living substances of any kind to occur in a fluid the organic constituents of which have been so altered by heat that they can retain no sort of chemical resemblance to the organic constituents of living matter. If the formation of life—of living substance—is possible at the present day—and for my own part I see no reason to doubt it—a boiled infusion of organic matter—and still less of inorganic matter—is the last place in which to look for it. Our mistrust of such evidence as has yet been brought forward need not, however, preclude us from admitting the possibility of the formation of living from non-living substance. "Presidential address to B.A.A.S." (1912), p. 12.

81. Ibid., p. 13.

NINE. FINAL ABANDONMENT

1. J. Keosian, *The origin of life* (New York: Reinhold Publishing Co., 1964), pp. 2–3. Keosian also asserts that mechanists believe organic matter to be solely a product of living organisms, while only materialistic hypotheses allow organic matter to be produced before organisms came into being. I see no validity in this distinction.

2. J. Keosian, "Life's beginnings—origin or evolution?," in J. Oro et al., eds., *Cosmochemical evolution and the origins of life* (Dordrecht, Holland: Reidel Publishing Co., 1974), p. 285.

3. J. D. Bernal, "The problem of stages in biopoesis," in F. Clark and R. L. Synge, eds., *Proceedings first international symposium on teh origin of life on the earth* (New York: Pergamon Press, 1959), p. 38. Since 1957 three other international conferences on the origin of life have been held: Wakulla Springs, United States, in 1963; Pont-à-Mousson, France, in 1970; and Barcelona, Spain, in 1973. The Barcelona meeting also represented the inaugural meeting of the International Society for the Study of the Origin of Life (ISSOL) and the birth of a new scientific journal, *Origins of Life*, published by D. Reidel.

4. H. Spencer, "On alleged 'spontaneous generation' and on the hypothesis of physiological units" (1870), reprinted as appendix to *Principles of biology*, vol. 1 (New York, 1898), p. 482.

5. H. Spencer, "Progress: Its law and cause" (1857), reprinted in *Essays: Scientific, political and speculative* (London, 1858), vol. 1.

6. See J. D. Burchfield, "Darwin and the dilemma of geological time," *Isis* 65

(1974): 301-21, and *Lord Kelvin and the age of the earth* (New York: Science History Publications, 1975).

7. For example, H. W. Mitchell, *The evolution of life* (New York, 1891), and E. Clodd, *Pioneers of evolution from Thales to Huxley* (London, 1897). H. Becquerel himself, in 1924, after criticizing Arrhenius's theory of panspermia, saw life originating at some stage in the earth's evolution.

8. P. C. Mitchell, "Abiogenesis," in *Encyclopaedia Britannica*, 11th ed. (Cambridge, 1910).

9. N. Lockyer, *Inorganic evolution as studied by spectrum analysis* (London, 1900), p. 159.

10. E. A. Schaefer, "Presidential address to B.A.A.S" (1912), p. 15.

11. Ibid., pp. 10-11.

12. R. Beutner, *Life's beginning on the earth.* (Baltimore: Williams and Wilkins, 1938), p. 83.

13. Ibid., p. 77.

14. Ibid., p. 68.

15. Keosian, *Origin of life*, p. 2.

16. B. Moore, *The origin and nature of life* (London: Thornton Butterworth, 1921), pp. 189-90.

17. Ibid., p. 190.

18. Keosian, *Origin of life*, p. 3.

19. A. I. Oparin, *The origin of life* (1936), English trans. S. Margulis (London: Macmillan, 1938; Dover reprint, 1953); all references to the Dover edition. Aleksandr Ivanovich Oparin (born 1894), a prominent Soviet biochemist, graduated in the natural sciences from Moscow University in 1917. From then until 1929 he specialized in biochemistry, working with the political revolutionary A. N. Bakh. In 1929, he was appointed to the chair of plant biochemistry at Moscow University. In 1935, he became associate director of the Bakh Institute of Biochemistry, Academy of Science.

20. The relationship between science and dialectical materialism in the Soviet Union has been discussed by Loren Graham in his brilliant and readable *Science and philosophy in the Soviet Union* (New York: Knopf, 1972). This wide-ranging study discusses Soviet interpretations of quantum theory, relativity theory, genetics, structural chemistry, cybernetics, physiology, and the origin of life. A detailed account of the dialectical basis of Oparin's work is included. See also *Marxism, communism and western society: A comparative encyclopaedia* C. D. Kernig, ed. (New York: Herder and Herder, 1972), especially articles on "Dialectical materialism" and "Life."

21. A. I. Oparin, *Life, its nature, origin and development,* trans. A. Synge (Edinburgh: Oliver and Boyd, 1961), p. 6.

22. F. Engels, *Dialectics of nature,* trans. C. Durt (New York: International Publishers, 1940).

23. This shift is discussed in Graham, *Science and philosophy in Soviet Union,* chap. 7, pp. 262-73.

24. Oparin, *Origin of life* (1936), p. 1.

25. Graham, *Science and philosophy in Soviet Union,* p. 17.

26. For details of the reappearance of the macromolecule concept, see J. T. Edsall, "Proteins as macromolecules: An essay on the development of the macromolecule concept and some of its vicissitudes," *Archiv. of Biochem. Biophysics* (Supplement 1) (1962): 12-20.

27. Oparin, *Origin of life* (1936), p. 27.

28. Ibid., p. 25.

29. Ibid., p. 33.

30. Ibid., p. 61.

31. These views were based on James Jean's catastrophic theory of planetary development, in which a fiery mass of the sun was torn out by the near passage of another star; the planet's subsequent history being one of gradual cooling. Also, work on the escape velocities of planetary atmospheres in the 1920s and 1930s had led to a general belief in an early reducing atmosphere of methane and ammonia. See H. N. Russell, "The atmospheres of the planets," *Science* 81 (1935): 1-9.

32. Oparin, *Origin of life* (1936), p. 109.

33. Ibid., p. 126.

34. Ibid., p. 137.

35. Ibid., pp. 146–48.

36. For an extensive discussion of coacervates by Bungenberg de Jong, see H. R. Kruyt, ed., *Colloid science*, vol. 2 (New York: American Elsevier, 1949), esp. p. 243.

37. Oparin, *Origin of life* (1936), p. 157.

38. Ibid., pp. 159–60.

39. Ibid., p. 160.

40. Ibid., p. 162.

41. Ibid., p. 174.

42. Ibid., p. 195.

43. G. Allen, *Life science in the twentieth century* (New York: John Wiley, 1975).

44. S. Miller, "A production of amino acids under possible primitive earth conditions," *Science* 117 (1953): 528. A list of other such experiments is given in S. W. Fox and K. Dose, *Molecular evolution and the origin of life* (San Francisco: W. H. Freeman, 1972), p. 128.

45. G. Wald, "The origin of life," *Scientific American*, August 1954, p. 1.

46. F. O. Schmitt, "Macromolecular interaction patterns in biological systems," *Proc. Amer. Phil. Soc.* 100 (1956): 476–86.

47. J. B. S. Haldane, "The origins of life," *New Biology* 16 (1954): 12–27.

48. J. B. S. Haldane was the only Marxist, to my knowledge, who favored the spontaneous generation argument. See ibid., and "Data needed for the blueprint of the first organism," in S. W. Fox, ed., *The origins of prebiological systems and their molecular matrices* (New York: Academic Press, 1965).

49. A. I. Oparin, *The origin of life on the earth,* 3rd ed. (Edinburgh: Oliver and Boyd, 1957), p. xi.

50. Z. A. Medvedev, *The rise and fall of T. D. Lysenko,* English trans. I. M. Lerner (New York: Columbia University Press, 1969).

51. Graham, *Science and philosophy in Soviet Union,* p. 275.

52. H. J. Muller, "The gene as the basis of life" (1926), in *Studies in genetics: The selected papers of H. J. Muller* (Bloomington: Indiana University Press, 1962), p. 198.

53. H. J. Muller, "The gene," *Proc. Roy. Soc. London* 134 (1947): 1–37; reprinted in E. A. Carlson, ed., *The modern concept of nature: Essays on theoretical biology and evolution by H. J. Muller* (Albany, N.Y.: State University of New York Press, 1973), p. 120.

54. H. J. Muller, "Life," *Science* 121 (1955): 1–9.

55. A. I. Oparin, "Forword," in Clark and Synge, eds., *First international symposium,* p. x. There were 46 scientists at the Moscow meeting from 16 countries. Approximately one-third of the papers were contributed by Soviet scientists.

56. W. M. Stanley, "On the nature of viruses, genes and life," ibid., p. 314.

57. N. H. Horowitz, "On defining life," ibid., p. 106–7. N. W. Pirie's claim was made in his article, "The meaninglessness of the terms life and living," in J. Needham and D. E. Green, eds., *Perspectives of biochemistry* (London: Cambridge University Press, 1938).

58. H. L. Fraenkel-Conrat and B. Singer, "The infective nucleic acid from tobacco mosaic virus," *First international symposium,* pp. 303–6.

59. E. Chargaff, "Nucleic acids as carriers of biological information," ibid., p. 299.

60. J. D. Bernal, "The scale of structural units in biopoesis," ibid., p. 386.

61. J. D. Bernal, "The problem of stages in biopoesis," ibid., p. 38. The English biochemist N. W. Pirie had coined the word earlier in "Ideas and assumptions about the origin of life," *Discovery* 14 (1953): 238. It had been also used in England, in an edition of *New Biology* 16 (1954) devoted to the origin of life question.

62. A. E. Braunshtein, discussion contribution, in *First international symposium,* p. 654.

63. N. I. Nuzhdin, "The qualitative specificity of living material," ibid., pp. 380–81.

64. Quoted in Medvedev, *Rise and fall of Lysenko,* p. 217.

65. See, in particular, *First international symposium,* pp. 368–80. All of the Soviet

contributors to this particular discussion opposed Stanley's views, and Engels' name was mentioned in three of the discussion papers. In addition, most of the Soviet participants were clearly arguing from a dialectical framework.

66. O. B. Lepeshinskaia, discussion contribution, in *First international symposium*, p. 484.

67. A. I. Oparin, discussion contribution, in ibid., p. 368.

68. Oparin, *Life,- its nature, origin and development*, p. 25.

69. H. J. Muller, "The gene material as the initiator and the organizing basis of life," *Amer. Nat.* 100 (1966): 500.

70. A. L. Lehninger, *Biochemistry* (New York: Worth Publishing Co., 1970), p. 782.

71. A. I. Oparin, "The appearance of life in the universe," in C. Ponnamperuma, ed., *Exobiology* (Amsterdam: North Holland Publishing Co., 1972), p. 3.

72. S. W. Fox, "A theory of macromolecular and cellular origins," *Nature* 205 (1965): 329. A similar positive attitude can be seen in his "Self-ordered polymers and propagative cell-like systems," *Die Naturwissenschaften* 56 (1969): 1–9. In this paper he waxed optimistic: "Questions that have been answered in principle include those of how enzymes arose in absence of enzymes to make them, how cells arose in the absence of cells to father them, how a first propagated cycle for enzymes and cellular structures could begin, and how macromolecular sequences with informational potentiality could arise from micromolecules, in the absence of nucleic acids." Fox also argued that the self-assembly of "proteinoid microspheres" was more applicable to the origin of life question than Oparin's coacervates. For details, see Fox and Dose, *Molecular evolution*, chaps. 6 and 7. This book remains one of the most lucid presentations of modern evolutionary views, although unfortunately the authors persist in using the phrase, "spontaneous generation as part of a theory of the origin of life."

73. H. F. Blum, *Time's arrow and evolution* (Princeton, N.J.: Princeton University Press, 1951).

74. A. I. Oparin, *The chemical origin of life*, English trans. A. Synge (Springfield, Ill.: Charles C Thomas, 1964), p. 107. Engels made a similar remark in his *Dialectics of nature*, p. 189.

75. Wald, "Origin of life," p. 6.

76. P. T. Mora, "The folly of probability," in S. Fox, ed., *The origins of prebiological systems* (New York: Academic Press, 1965), p. 39. Ludwig von Bertalanffy uses the same argument; see Kernig, ed., *Marxism, communism*, vol. 5, p. 218. He mentions also that acceptance of the spontaneous generation viewpoint depends upon extra-scientific factors, such as "the pecuniary interests of space programmes." He was obviously referring to support given by the biological section of the NASA space program to American biochemists studying the origin of life.

77. Wald, "Origins of life," p. 6.

78. Ibid., p. 4.

79. A. S. Konikova, discussion contribution, in *First international symposium*, p. 117.

80. Fox and Dose, *Molecular evolution*, p. 13.

81. A. I. Oparin and V. Fesenkov, *Life in the universe*, trans. D. A. Myshne (New York: Twayne Publishers, 1961), p. 16.

82. A. I. Oparin, *Genesis and evolutionary development of life* (New York: Academic Press, 1968).

83. Oparin, "Appearance of life."

84. J. Keosian, "On the origin of life," *Science* 131 (1960): 479–82 and "Life's beginnings—origin or evolution?" in Oro et al., eds., *Cosmochemical evolution*.

85. A. E. Smith and D. H. Kenyon, "Is life originating de novo?" *Perspectives in biology and medicine* 15 (1972): 529–42.

INDEX

THE JOHNS HOPKINS UNIVERSITY PRESS

This book was composed in VIP Baskerville text
and display type by The Composing Room of Michigan
from a design by Susan Bishop. It was printed on
50-lb. Publishers Eggshell Wove and bound in Joanna
Arrestox cloth by The Maple Press Company.

LIBRARY OF CONGRESS CATALOGING IN PUBLICATION DATA

Farley, John, 1936–
    The spontaneous generation controversy from Descartes to
Oparin.

    Includes index.
    1. Spontaneous generation.   I. Title.
QH325.F24   577   76-47379
ISBN 0-8018-1902-4